Immunological Aspects of Mammalian Reproduction

Immunological Aspects of Mammalian Reproduction

PETER J. HOGARTH, B.Sc., D.Phil.

Lecturer in Biology
University of York

Blackie

Glasgow and London

Blackie & Son Limited
Bishopbriggs
Glasgow G64 2NZ

Furnival House
14/18 High Holborn
London WC1V 6BX

First published 1982

British Library Cataloguing in Publication Data

Hogarth, Peter J.
Immunological aspects of mammalian reproduction.
1. Reproduction—Immunological aspects
2. Mammals—Reproduction
I. Title
599.01′66 QP251
ISBN 0-216-91247-4
ISBN 0-216-91248-2 Pbk

Filmset by Advanced Filmsetters (Glasgow) Ltd
Printed by Thomson Litho, East Kilbride, Scotland

Preface

Until quite recently, the field of reproductive immunology was very much a neglected area of biology, seen by most reproductive physiologists as of only peripheral importance. It was generally acknowledged as curious that a female mammal tolerated the intrusion of alien sperm and the persisting presence of an alien fetus, while reserving the prerogative of rejecting grafts of tissue, even when these were from her own mate. Several theories were advanced to explain this paradox, each with some supporting evidence: all were eventually shown to be inadequate. And there the matter was, on the whole, permitted to rest.

In the last few years, the situation has changed dramatically, and the neglected area of overlap between immunology and reproduction has again become densely populated by research workers. As a symptom of this resurgence of interest, a specialist journal (the *Journal of Reproductive Immunology*) has been launched to supply what had rapidly been perceived as a need.

The reasons for this transformation are various. Immunology itself has entered a new and exciting phase, in which many fundamental problems are at last yielding to investigation. More sophisticated techniques have been developed in which the power and precision of the immune system have been exploited, and gene manipulation and the production of monoclonal antibodies represent a further refinement whose impact over a range of fields (not least in reproduction) promises to rival that of the original invention of histological stains. The newly achieved perceptions of the nature of the immune system and the powerful practical applications of immunological techniques, when brought to bear on reproduction, have established that the whole question of the immunological relationship between mother and fetus is one of fascinating richness and subtlety.

Finally, various forms of infertility and reproductive disorder in humans and domestic animals have been shown to be affected by immunological considerations, which has given an added incentive to investigators.

Reproductive immunology is therefore no longer a fringe interest, a small and remote subject of which biologists know little, but an exciting and burgeoning area of research, of great intrinsic interest and considerable practical significance.

In this book I attempt to explain some of the subtlety and complexity of the field and to discuss some of the current ferment of ideas. I hope it will stimulate a wide range of people with an interest in biology: perhaps reproductive physiologists and medical scientists primarily, and advanced undergraduates and postgraduates in those areas. These are the main readers for whom I have written, but I hope that a subject of such fascination and importance will find a wide audience among biologists.

Many people helped in various ways with the production of this book. I should particularly like to record my appreciation and thanks to Dick Hunter for his excellent and painstaking work on the illustrations.

P.J.H.

Contents

CHAPTER ONE

INTRODUCTION

THE IMMUNE SYSTEM OF MAMMALS IS EXTREMELY EFFECTIVE IN DISPOSING OF foreign materials. A graft of skin from an unrelated donor is rejected typically in ten days or so; a foreign protein injected in solution would be completely cleared from the circulation of the recipient in about the same period. In both cases a state of immunity is induced, and persists for a considerable time, possibly for the rest of the immunized animal's life. Every subsequent presentation of the same foreign material will be dealt with even more rapidly and effectively than the first.

Much the same is true of the majority of microorganisms. Bacteria, viruses, and fungal infections are usually dealt with in a matter of days or weeks. Although they may sometimes survive, and cause symptoms of disease for longer than this (or even prove fatal), the vast majority of pathogens or potential pathogens are speedily eliminated, or at least restrained to tolerable levels, by the immune system.

Reproductive processes present several circumstances where the immune system notably fails to destroy foreign material when it might be expected to do so. Firstly, during copulation, spermatozoa are deposited within the genital tract of the female: yet, despite their paternal (that is, 'foreign') origin, they are not destroyed (or at least not until fertilization has probably occurred, when it no longer matters). A second insemination with spermatozoa from the same source is no less likely than the first to succeed in fertilization; if immunity to spermatozoa (or to some other constituent of semen) does develop and persist, it is singularly ineffective.

Pregnancy is, of course, the second obvious circumstance in which foreign material survives and the immune system apparently declines to function normally. The fetus is, at least superficially, analogous to a graft of skin. Genetically it is half paternal in origin, and by virtue of this partly paternal nature, should be rejected as a conventional skin graft from the father would be. This does not happen: the fetus develops and sustains an intimate relationship with the mother for the duration of pregnancy, which lasts in some species for many months.

Finally, although the relationship after birth between the suckling young and its mother is less intimate than that during pregnancy, there are nevertheless some interesting immunological implications.

This book is devoted to exploring some of the immunological implications of mammalian reproduction, and the ways in which potential ill effects of the immune system are avoided so that the initiation, continuation, and eventual success of pregnancy are ensured.

CHAPTER TWO

THE IMMUNE SYSTEM

2.1 Humoral immunity

If a mouse is injected subcutaneously with a suspension of appropriate bacteria, two outcomes are possible. Either the mouse falls ill and succumbs to the infection within a few days, or it recovers. In the latter case, it is usually found to be resistant to any subsequent attempts at reinfection with the same type of bacterium; it has become *immune*.

The course of the infection can be monitored by obtaining daily blood samples and assessing the bacterial content of the blood. Figure 2.1 shows what might occur. Initially, the concentration of bacteria is determined by the number injected and the volume of the mouse in which they are dispersed. As the bacteria multiply, their concentration in the blood rises. If it continues to do so, the mouse will die; if the situation can be controlled, the multiplication of the bacteria is curbed, their numbers dwindle, and after 10 to 14 days none remain. A second injection a few weeks later is summarily disposed of: in this case it is likely that the infected mouse will scarcely display any symptoms of disease.

This experiment demonstrates one crucial aspect of the immune response: a form of immunological memory is present, so that the mouse 'remembered' its exposure to the given stimulus and reacted more rapidly and effectively to a second stimulus. The state of induced immunity will persist for a long time, probably indefinitely. The immune system is thus an adaptive one, in that its activities are adapted to the prevailing threats.

A second important aspect is seen if our immunized mouse is injected with bacteria of a different type. The response to these is of the primary type, with clearance of the bacteria in 10 to 14 days. The immune system is therefore discriminating; even if two types of bacteria are very closely related, they can be distinguished immunologically. In some circumstances, immune mechanisms can tell apart even such similar molecules as proteins

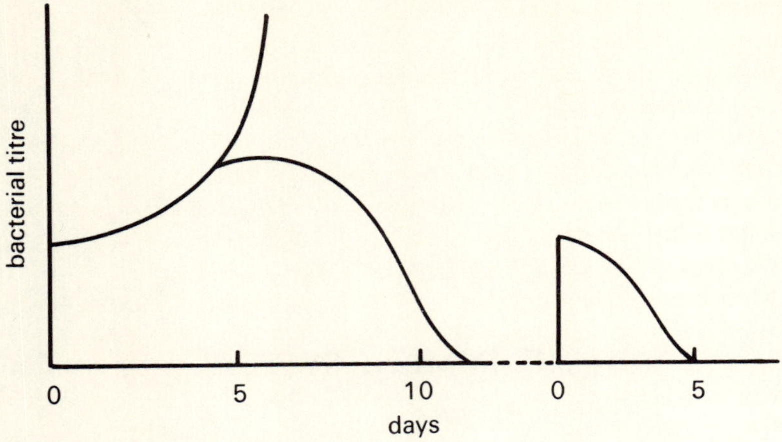

Figure 2.1 Bacterial concentrations in the blood of a mouse at intervals after the injection of live bacteria of a strain never before encountered (left), and a second injection of bacteria of the same strain (right). Following the first injection, the mouse either succumbs to the infection or recovers and eliminates bacteria within 10–14 days; after a second injection, elimination is more rapid, demonstrating the difference between primary and secondary responses.

differing by only a single amino acid, or the different optical isomers of such low molecular weight substances as tartaric acid. There are exceptions—the immunological pun which enables immunization against the cowpox virus to confer immunity against smallpox also is one which has proved useful—but in general the immune system is highly specific in its operations.

Stimuli which can provoke an immune reaction are described as *antigenic*. Bacteria and other cells are antigenic by virtue of surface molecules of various types, and the molecular structures which endow a cell with antigenicity are obviously of interest. Generally, proteins of molecular weight greater than 5 000 or so are antigenic. There are exceptions, including synthetic polypeptides of low molecular weight which are antigenic, and some large proteins such as collagen, with highly repetitive amino acid sequences which seem to have insufficient variety to arouse the interest of the immune system. Polysaccharides of high molecular weight may be antigenic; the specific elements of the blood group antigens of the human ABO system, for instance, are carbohydrate. Lipids and nucleic acids usually make poor antigens.

A further complication is that many substances lack the ability to induce antibody formation, but may combine with specific antibody that is already formed. These are known as *haptens*: a hapten coupled to a carrier protein which is not itself a good antigen may cause an immune response directed not against the carrier substance, nor even the complex, but against the hapten itself.

How are the bacteria, or other antigens such as injected protein solution, destroyed? An important role is played by specific globulin proteins known as *antibodies*, or *immunoglobulins.* The bulk of these fall into the γ-globulin category and in consequence 'antibody' and 'γ-globulin' are sometimes used as interchangeable terms. However, as not all antibody belongs to the γ-globulin fraction of blood serum, the terms are not in fact synonyms.

Specific antibody appears in the blood serum within a few days of the antigenic stimulation, and rises to a plateau level. After removal of the bacteria, the antibody may remain at detectable levels for some time, but usually declines and disappears. When a secondary response is evoked, the antibody appears sooner after the stimulus and rises steeply, probably to a higher level than before. The course of primary and secondary responses may thus be described in terms of antibody concentration, as well as of antigen concentration (Figure 2.2).

How is antibody produced? When antigen is injected into the skin, it is first drained off in tissue fluid and passes through minute lymphatic vessels into a regional lymph node (Figure 2.3). The regional lymph nodes are small encapsulated structures situated at strategic points throughout the body, each having a certain constituency of tissue from which it draws the fluid. The spleen, among other functions, acts as a lymph node whose constituency is the blood circulation as a whole. Other lymphoid tissues within the body include the bone marrow and thymus, as well as patches in the epithelia of the genital tract and small intestine, the tonsils, and the vermiform appendix.

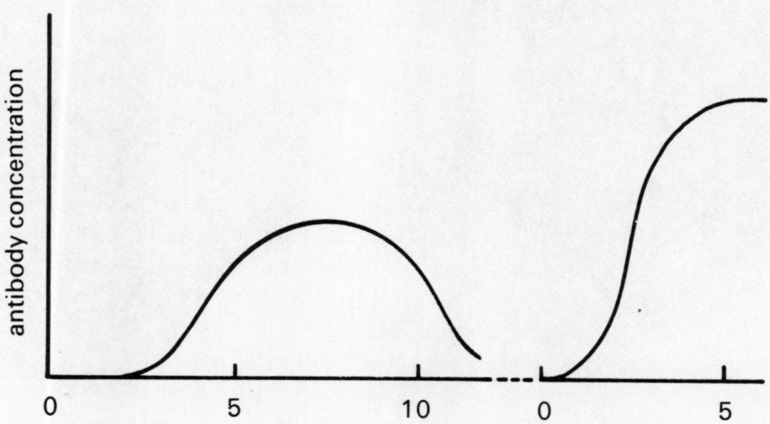

Figure 2.2 As Figure 2.1; here the concentration of antibody against the injected bacteria is plotted.

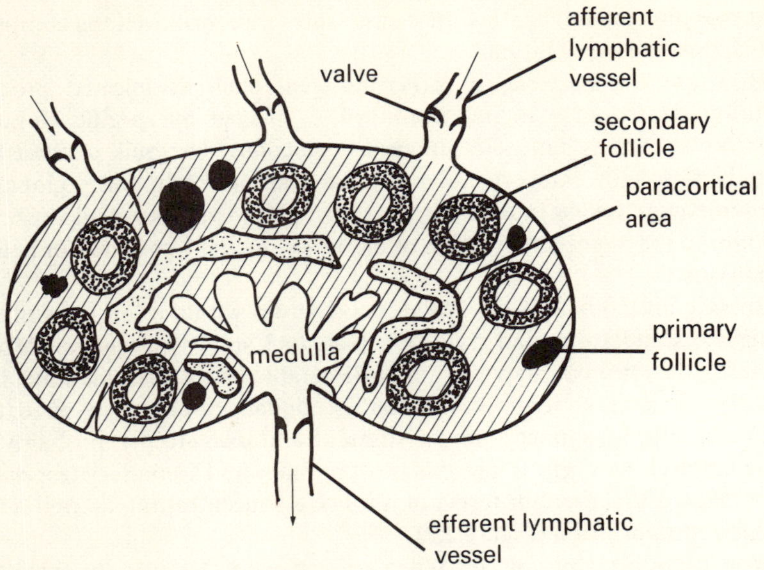

Figure 2.3 Diagrammatic representation of a regional lymph node. The cortex consists largely of B lymphocytes interspersed with macrophages. Before stimulation with antibody these form dense primary nodules or follicles; after stimulation the secondary follicles appear as concentrically packed aggregates of small lymphocytes surrounding paler staining germinal centres. T cells are predominantly restricted to the paracortical area of the node.

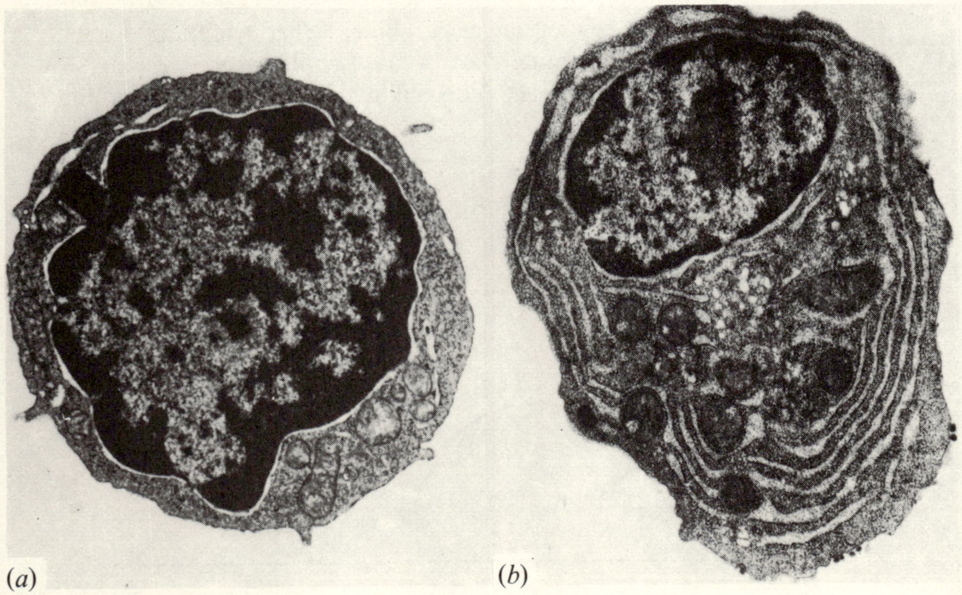

Figure 2.4 Cells involved in the immune response: (*a*) a lymphocyte; (*b*) a plasma cell (×10000). (From Bussard, A. (1973) L'origine cellularie des anticorps. *La Recherche* **4** 31: 115–24, Figs. 7*b*, 7*c*, p. 122.)

Within each regional lymph node is a tightly-packed mass of cells, most of which are classified as *lymphocytes* (Figure 2.4). A typical lymphocyte is a small cell with a relatively large nucleus and rather little cytoplasm. When antigen reaches a lymph node, a small fraction of the lymphocytes respond by starting to divide rapidly. In the course of this proliferation, some develop more cytoplasm, containing much rough endoplasmic reticulum; these are known as *plasma cells*. Rough endoplasmic reticulum is a sign of the active synthesis of protein, making the plasma cells obvious candidates for producers of antibody. This view has been confirmed by the investigation of single plasma cells *in vitro*. A point of great theoretical interest is that an individual plasma cell makes antibody of only a single specificity. It is therefore fully committed as to its future target antigen by, at the latest, the period of proliferation.

As the primary response subsides, plasma cells disappear again. In the secondary response they reappear rapidly, in strength, presumably developing from the many committed lymphocytes left after the proliferative phase of the previous primary response. A primary response therefore has the effect of picking out a particular subset of lymphocytes in the body, defined by having a particular specificity, and increasing their numbers by proliferation. The 'memory' lymphocytes which persist after the primary response form the basis of the more intense secondary response.

Antibody, and some of the stimulated lymphocytes, are released from the regional lymph node where the response started, and pass through lymphatic vessels, eventually into the blood circulation (Figure 2.5). In this

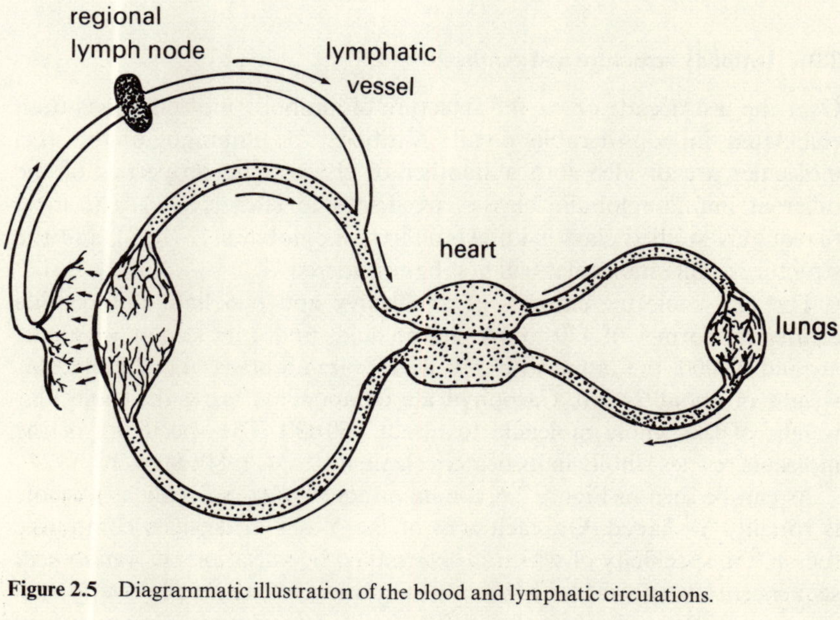

Figure 2.5 Diagrammatic illustration of the blood and lymphatic circulations.

way the immune response is spread throughout the body, with activated lymphocytes and plasma cells soon cropping up in other lymph nodes, and antibody in the blood serum finding its way to all parts of the body served by blood capillaries.

When a particular antibody molecule encounters a molecule of the antigen against which it is directed, it combines with it. If the antigen is a soluble one such as a protein, the antigen–antibody complex is then ingested by a phagocytic cell. In the case of a bacterial cell, the attachment of antibody molecules to the surface might be an embarrassment but would scarcely, on its own, prove rapidly lethal. The killing of bacteria is carried out partly by a further action supplied by substances known collectively as *complement* (C′). Complement actually comprises nine protein components (numbered C1 to C9, with the first component divided into three subfractions). Normally, complement circulates in solution in the blood serum. When antibody molecules attach to a target cell surface in sufficient density, the complement components recognize the presence of attached antibody, and sequentially assemble themselves on it to form a complex structure which has the ability to penetrate the bacterial surface. In the process of complement self-assembly, phagocytes are attracted into the area. The result is lysis of the target cell and rapid removal, by phagocytosis, of the debris. Recognition of the foreign cell is therefore specific (by antibody attachment); the actual destruction is non-specific, by molecules which could have become involved in any other complement-mediated activity (Burnet, 1976; Inchley, 1981; Playfair, 1979; Roitt, 1980).

2.2 Antibody structure and synthesis

Over the last decade or so, the structure of antibody molecules has been elucidated in considerable detail. Antibody, or immunoglobulin (Ig) molecules are divided into a number of classes; the properties of the different immunoglobulin classes are discussed later (p. 11). The most thoroughly studied class is known as immunoglobulin G (IgG), and the structure of IgG molecules will first be considered.

The IgG molecule consists of two heavy and two light polypeptide chains, the former of 420 to 440 amino acids and a molecular weight of around 48 000, the latter with 210 to 230 amino acids and a molecular weight of about 24 000. Carbohydrate components bring the molecular weight of the whole molecule to about 150 000. The specificity of the molecule resides wholly in its protein element (Roitt, 1980; Steward, 1974).

As can be seen in Figure 2.6, the immunoglobulin molecule as a whole is roughly Y-shaped. On each arm of the Y lies an antigen-combining region, the specificity of which is determined by variations in amino acid sequence from one antibody species to another. Variations in amino acid

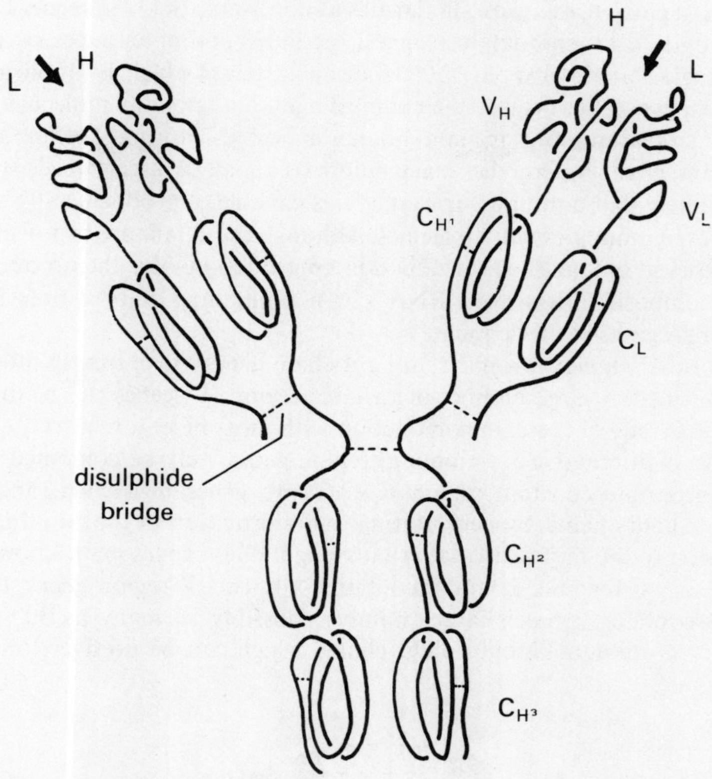

Figure 2.6 Diagrammatic representation of an immunoglobulin (IgG) molecule. (H, L = heavy, light chains; V_L, V_H = variable domains; C_{H^1} etc. = constant domains; arrows indicate antigen-binding sites.)

sequence affect both heavy (H) and light (L) chains. Outside the variable (V) region, both heavy and light chains are relatively, although not absolutely, constant in amino acid sequence. This is why, when the 'tails' of a mixed bag of immunoglobulin molecules (such as would normally be found in blood serum) are removed by digestion with the enzyme papain, they can be crystallized: only molecules which are identical in structure, or nearly so, can assemble into a regular crystal. In contrast to the crystallizable fragment (Fc) of the immunoglobulin molecule, the variable antigen-binding (Fab) fragments released by papain are not crystallizable since these differ from molecule to molecule.

A pattern of folding characteristic of immunoglobulin molecules repeats itself four times within each heavy chain, and twice within each light chain, so that the chains can be divided into four (three constant, one variable) and two (one constant, one variable) domains, respectively. Within each constant domain (numbered C_{H^1} to C_{H^3} and C_L), besides the similarity of folding pattern, there are similarities also in amino acid sequence. These homologies between domains suggest the likelihood of an ancestral gene for a single domain of about 12 000 molecular weight which, by duplication and divergence, has produced a composite immunoglobulin molecule.

One crucial question in immunology is how the diversity of immunoglobulins is generated; the exact number cannot be ascertained, but it seems likely that a mouse, for example, is capable of producing 10^8 or so different immunoglobulin molecules. Although the details are by no means fully worked out, antibody production appears to involve the production of a composite messenger RNA chain containing information from different regions of the genome.

Figure 2.7 shows how different light chain genes are currently thought to be formed. A large number of variable region (V) genes can be drawn upon. Any one of these, in conjunction with any one of a relatively small number of alternative J (=joining) region genes, may be combined with an appropriate constant region (C) gene to generate a wide range of possible light chains, by permutating the information in the initial DNA sequence. In the mouse, two alternative light chain genes exist, known as κ and λ; whereas the λ family contains only two V region genes, the κ family contains a much larger number, possibly as many as 300. The number of immunoglobulin light chains which can be produced in this

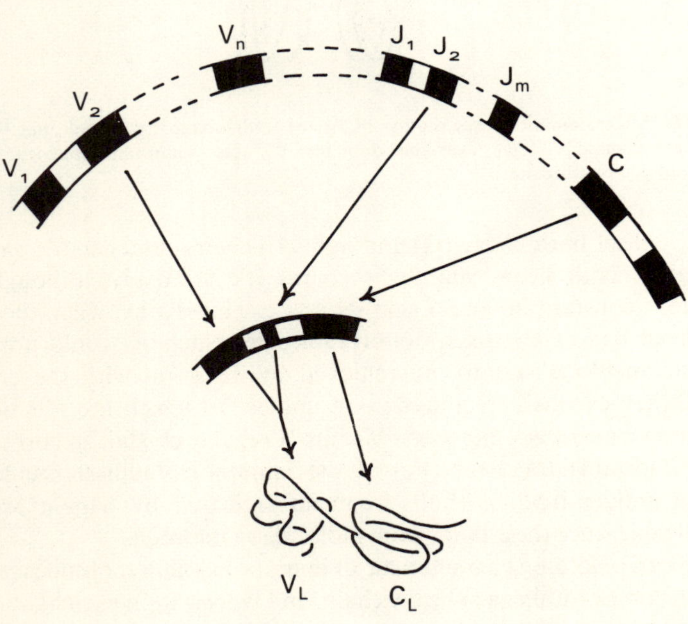

Figure 2.7 Genetic specification of the structure of the mouse (light) chain. (V_l–V_n are V-region genes, J_l–J_m are joining region genes and C specifies the constant domain.)

way is high, and is increased still further by some permitted variation in the exact site of joining of V and J genes.

With heavy chains, a further twist in the plot is introduced by a D (=diversity) region interpolated between the V and J elements. The D region introduces considerable additional diversity into heavy chain production by a mechanism whose elucidation is at only a preliminary stage. In view of this, and since the proposed mechanism is 'of such baroque complexity that it would never have occurred to anyone to suggest it but for the evidence...in the DNA' (Robertson, 1981) perhaps the eccentricities of the D region can be passed over hurriedly.

Each chain of the IgG molecule is therefore a composite, coded for by stretches of DNA in different and non-adjacent regions of the original genome. The final immunoglobulin molecule is formed by combining two identical heavy chains with two identical light chains; different combinations of heavy with light chains will generate yet further variants.

These genetic rearrangements occur randomly, and well before the cells involved are capable of actually producing antibody. The effect is to generate numerous clones of lymphocytes, members of each clone being committed in advance to producing antibody of a particular specificity. Commitment precedes exposure to relevant antigen; many clones may never, in the event, encounter the antigen to which rearrangement of their immunoglobulin genes has committed them.

After final commitment has been made, before a lymphocyte matures, as to the specificity of the antibody which it can produce, the same variable region may be combined with an alternative constant region to produce antibody belonging to a different class. The properties of immunoglobulins of the major classes are described in Table 2.1. The basic unit is very similar to that of IgG; the differences between classes are based partly on the incorporation of different heavy chain constant regions, and partly on the tendency of some immunoglobulins to exist as multiples of the basic four-chain molecule; IgM, for example, is generally found in the blood serum as a pentamer. Immunoglobulins of the IgG class can be further subdivided into four subclasses, each of somewhat differing properties. Human IgG4, for instance, binds complement less well than IgG1; it might therefore in some circumstances protect foreign cells from complement-mediated attack by competing for the same antigenic sites as IgG1 (see p. 22) (Inchley, 1981; Playfair, 1979; Roitt, 1980).

IgA is the principal immunoglobulin of mucous secretions. It is secreted as a dimer consisting of two IgG-like molecules joined by a cysteine-rich polypeptide of molecular weight 15 000, the J-chain (not to be confused with the J-region of IgG; see above). During transport and secretion this molecule is further combined with a 'secretory piece', a polypeptide of molecular weight 60 000, which plays some part in accumulating IgA out

Table 2.1 Properties of the immunoglobulin classes

Class	*Molecular weight*	*Heavy chains**	*No. of units*	*No. of combining sites*	*% of total serum Ig*	*Complement fixation*	*Principal function*
IgG	150 000	γ	1	2	80	+ +	combats microorganisms and their toxins
IgA	160 000 and 320 000	α	1 and 2	2 and 4	13	–	major Ig of mucous secretions; external defence
IgM	900 000	μ	5	10	6	+ + +	effective agglutinator, e.g. of bacteria in blood
IgD	185 000	δ	1	2	0–1	–	almost all on lymphocyte surface; disappears during proliferation or switch from IgM to IgG; a receptor, primarily?
IgE	200 000	ε	1	2	0.002	–	protects external surfaces; combating parasite infections; responsible for allergy symptoms

* All light chains may be either κ or λ.

of blood serum, transporting it across epithelial cells, and protecting it against proteolytic digestion (Bienenstock and Befus, 1980).

As a lymphocyte matures and becomes capable of responding to antigen and conversion into a plasma cell, it produces small amounts of antibody and displays this on its surface membrane with the antigen-binding (Fab) sites exposed (Robertson, 1980). This initial antibody appears to be a monomeric form of IgM and serves as a surface receptor for a specific antigenic determinant. It is of course possible that no antigen with this determinant will ever arrive. If it does, then the surface antibody binds it, and the antibody-bearing lymphocyte is stimulated to proliferate. Many of its descendants become plasma cells and secrete large amounts of antibody of the same specificity as the surface receptor antibody: at first, pentameric IgM and later IgG of the same specificity as the IgM. The same V-, J- and D-region genes are now incorporated with a different set of constant-region genes. Lymphocytes which lie within mucosal epithelia (including those within the lining of the genital tract; see p. 55) produce IgA in place of IgG (Bienenstock and Befus, 1980). In either case the antibody is released into the surroundings in large amounts and an effective soluble, or humoral, antibody response is under way.

2.3 Cell-mediated immunity

In the case of a tissue graft, the antigens which elicit rejection remain attached to the surface membranes of graft cells; they cannot therefore reach lymph nodes in soluble or suspended form. A somewhat different mechanism is needed to detect foreign grafts and take steps to destroy them.

Before embarking on an account of the mechanisms of graft rejection, it is important to clarify the terminology which is applied to tissue grafts. A graft from a genetically different member of the same species is referred to as an *allogeneic* graft, or *allograft*. *Xenografts* (or *heterografts*) come from members of another species, *autografts* from the recipient itself, and *isografts* from a different, but syngeneic, individual; for example, from another member of the same inbred, and consequently genetically homogeneous, strain of mouse.

The cells involved in graft rejection are also lymphocytes, but of a different subset from that to which the lymphocyte precursors of plasma cells belong, and are known as T lymphocytes. Whereas all mature lymphocytes stem from precursors within the bone marrow, those which develop into T cells spend some time in the thymus (hence ‘T’) before colonizing the regional lymph nodes and other peripheral lymphoid tissues. Cells which do not enter the thymus before circulating through the peripheral lymphoid tissues are termed B lymphocytes, and it is these which are responsible for the soluble antibody responses already described.

The designation of the cells of the soluble, or humoral, antibody response as B cells stems from the finding that, in birds, those cells must pass through a specific organ, the bursa of Fabricius, to become fully functional, just as the lymphocytes involved in graft rejection responses must pass through the thymus. No equivalent of the bursa of Fabricius has been identified in mammals and, in fact, there is evidence that commitment to a career as a B cell takes place before the future B cell even leaves the bone marrow. Fortunately, however, B stands for bone marrow as well as bursa, so the terminology remains reasonably apt (Solomon and Horton, 1977).

As well as becoming committed to graft rejection, under the influence of the thymus the T lymphocytes of mice develop a convenient surface marker by which they can thenceforth be recognized. As this is identifiable only by immunological methods, it is referred to as the Thy-1 antigen. Other species have similar surface markers which can be used to distinguish T from B cells, but on a less straightforward basis. Typically, some 70% of peripheral lymphocytes are classifiable as T cells on the basis of their surface markers, and 10–20% as B cells; leaving a certain amount of leeway to accommodate related cells which can be allocated to neither category.

A proportion of T cells carry surface receptors for the Fc portion of IgG or IgM molecules, and may in consequence also carry small amounts of adsorbed surface antibody. Attachment by the Fc portion leaves the specific antigen-binding (Fab) portions exposed, and it was formerly thought that a T cell would become primed by the passive acquisition of surface antibody synthesized by B cells. As it travelled round the body, it might then eventually come into contact with the target antigen carried by a graft cell; combination of antigen with antibody would constitute recognition of the 'foreignness' of the target, and destruction would ensue.

This is now known not to be the case, although it is not very far from the truth. T cells *do* carry surface receptors, specific to particular antigens, but these are self-generated. They are not immunoglobulin molecules, although similar in structure to immunoglobulins. The T cell antigen receptor appears to consist of two chains, each the product of an immunoglobulin heavy chain V-region gene, spliced to a constant region which is not one of those used in manufacturing immunoglobulins (Ohno *et al.*, 1980). Presumably the diversity of T cell specificities, and commitment to these, is generated in a similar way to the generation of diversity among the immunoglobulin molecules which serve as surface receptors on B cells, and the initial combination of antigen with receptor similarly induces a proliferative response.

The progeny of the stimulated T cell must include memory cells to ensure a more potent secondary response, but no plasma cells form from proliferating T cells. Instead, the effector T cells must find their way back to the graft site and act directly to kill graft cells, not by proxy.

Killing of graft cells is not the only operation performed by T cells; they also interact with B cells in various complex ways. A fully effective humoral antibody response will not generally occur without the participation of helper T cells (although B cell responses to some antigens are independent of T cell help). In addition to specific helper T cells, specific suppressor T cells may also arise; these suppress a B cell response to a particular antigen, in most cases, but may also have a more general inhibitory effect. Suppressor and helper T cells are distinguishable subsets of the T cell population, and can be discriminated from each other, and from cytotoxic T cells, by surface marker molecules.

A further subset of T cells is concerned with the phenomenon of delayed hypersensitivity, which includes a number of allergic types of response. These cells respond to stimuli with the production of lymphokines, substances which act on macrophages and other lymphocytes, and incite them to participate in various forms of aggressive action against target cells (p. 24). Targets are generally either allograft cells or endogenous virus-infected cells.

There are therefore a number of ways in which T cells interact with other T cells, and indeed this may be the primary role of helper and suppressor cells. Helper T cells facilitate the emergence of suppressor cells from their precursors: suppressor T cells inhibit helper cells. This provides the basis of a possible immunoregulatory system, in which the nature of the immune response which develops in a particular set of circumstances, and its effect on a target allograft, may depend on the balance between suppressor and helper activity. Interaction of the cells of the immune system with foreign antigen is then a function of their interactions with each other (Pernis and Vogel, 1980; Reinherz and Schlossman, 1980).

Different kinds of T and B cells are not the only cells which take part in alloantigen destruction. The initial filtering of a soluble antigen from the tissue fluid is carried out by macrophages in the lymph node (see Figures 2.3 and 2.4), and the development of an immune response depends on antigen being presented to a lymphocyte by a macrophage. The lymphocyte, to respond, must simultaneously recognize both the antigen and the macrophage. Macrophages are also involved in target cell destruction, as phagocytes. They carry high affinity receptors for the Fc region of IgG, as well as for complement component C3b; both forms of receptor will help to identify a cell to which complement-fixing antibody is attaching as a suitable candidate for attack (Rosenthal *et al.*, 1978).

In addition to macrophages, a variety of other related cells may be involved in rejection responses in various specialized capacities. *Killer* and *natural killer* (NK) cells have been identified (Marx, 1980), although their roles are not yet clear; other cells such as eosinophils are involved in various ways in dealing with various forms of threat. The relationships of these cell types are shown in Figure 2.8. Their precise roles are at present conjectural, and as they play relatively little part in the immunology of reproduction, they need not be discussed further.

2.4 Histocompatibility antigens

The antigens of allogeneic cells to which T cells respond are termed transplantation or histocompatibility antigens. As almost any protein can be antigenic, many cell surface components are potential candidates; providing only that they are sufficiently polymorphic within a species to ensure the probability of recognizable differences between graft donors and recipients. Only 'non-self' antigens are normally recognized as foreign;

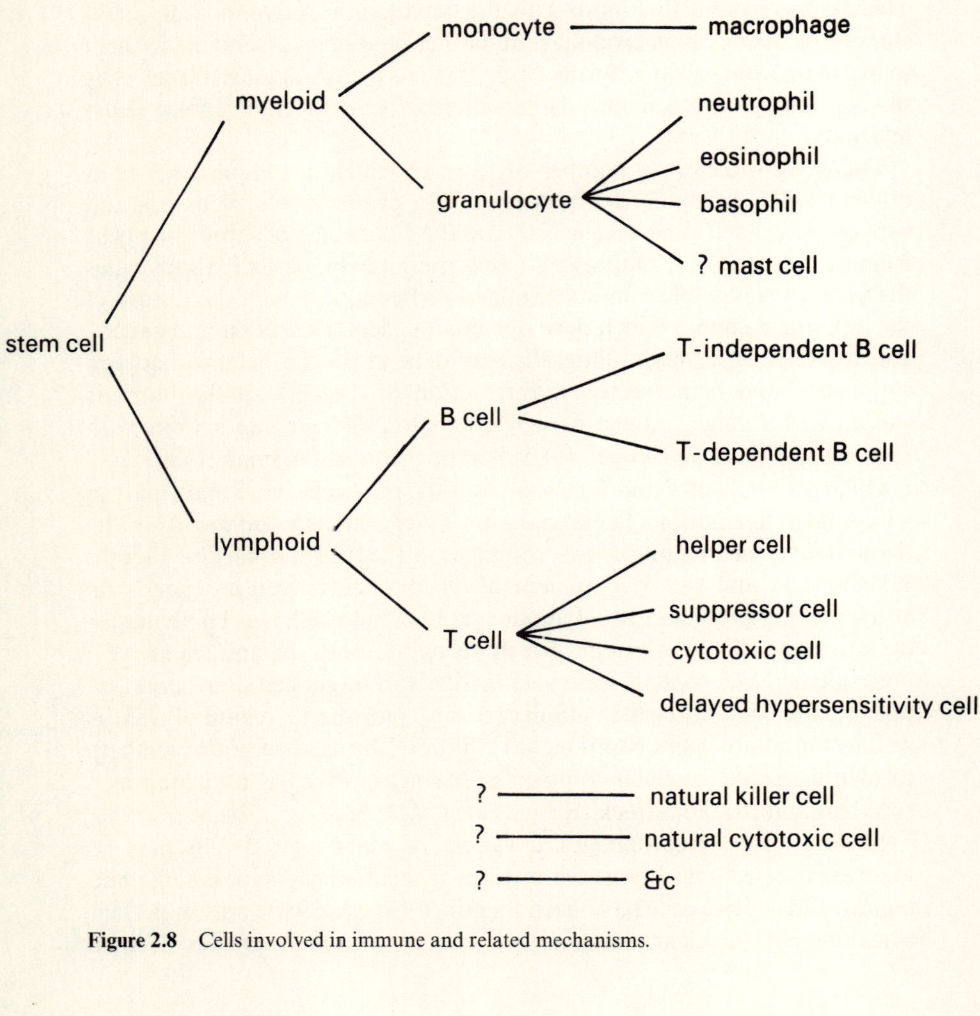

Figure 2.8 Cells involved in immune and related mechanisms.

'self' equivalents are not. *Autoimmune* reactions do, however, sometimes occur, and autoantibody formation is a feature of many human diseases including rheumatoid arthritis, pernicious anaemia and some forms of diabetes. Spontaneous antibody formation to a number of autoantigens has also been shown in the absence of any apparent ill effects (Golub, 1980; Roitt, 1980; Shulman, 1974).

In practice, allografts are destroyed because of a response to a particular subset of cell surface components. These appear to have no function other than in the initiation of an immune response by allogeneic T cells. If a graft donor and host differ with respect to such a major histocompatibility antigen, rejection will be prompt. If donor and host are identical with respect to their major histocompatibility antigens, the residual incompatibility of minor antigens will probably not suffice for rapid graft rejection; survival will be prolonged, possibly indefinitely.

In the mouse, the genetic region determining the major histocompatibility antigens is designated H-2, and the minor loci H-1, H-3, and so on, up to a total of about 30 loci relevant to graft rejection. In humans, the major region is termed HLA; again, many minor loci probably exist (Amos and Kostyu, 1980; Iványi, 1978; Klein *et al.*, 1981).

Clearly, the major histocompatibility antigens did not evolve solely to complicate the lives of transplant surgeons and to intrigue biologists, although both of these results are undeniable. Almost certainly, their primary function is in facilitating the response to intracellular virus infections. By their nature, viruses spend much of their time within host cells, not free in the blood or other body fluids. While actually inside cells, they are not readily accessible to the humoral antibody system. A major role of cell-mediated immunity is to recognize and eliminate virus-infected cells.

Major histocompatibility antigens play an essential role in the process of recognition. Cytotoxic T cells can recognize viral antigen at the surfaces of infected cells, but only in conjunction with the intrinsic major histocompatibility antigens; and then only if the antigens borne by the target cell are similar to its own. Lymphocytes of a mouse of one genetic strain can recognize and kill virus-infected cells of the same strain; they are unable, in general, to destroy similar cells from a mouse of a different strain, even if these are infected with virus of identical type. It is not clear whether viral and histocompatibility antigens are recognized independently but simultaneously, or whether they interact with each other and are recognized as a joint entity. Whichever is the case, the capacity to recognize intracellular infection is restricted by the antigens borne (Howard, 1980; Robertson, 1979).

The role of histocompatibility antigens in allograft recognition and destruction is therefore a secondary result of their primary involvement in combating certain viral infections.

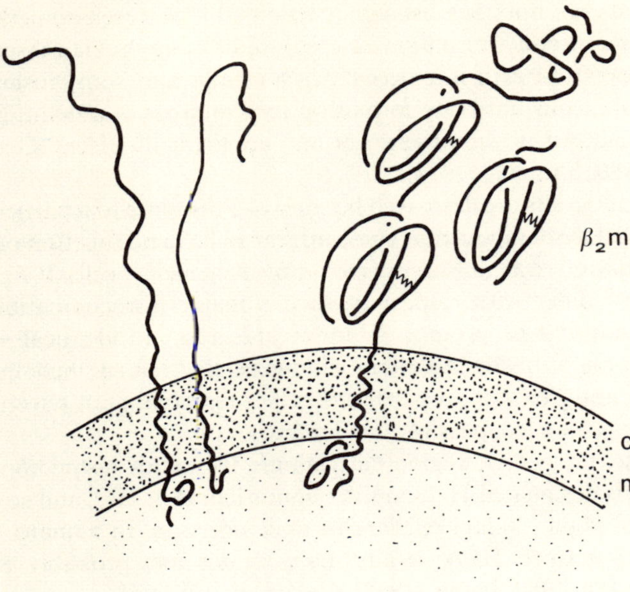

Figure 2.9 Structure of products of the Major Histocompatibility Complex (MHC). On the right is a classical histocompatibility antigen (after Strominger, 1980); on the left an Ia antigen. For further description, see text. (β_2m = β_2-microglobulin.)

The structure of a human HLA transplantation antigen molecule is shown in Figure 2.9. It bears a more than coincidental resemblance to an immunoglobulin molecule (Figure 2.6). A heavy chain, of molecular weight 43 000, is associated with a light chain of molecular weight about 12 000. The region of the heavy chain nearer to the point of insertion in the cell membrane is relatively constant in structure between individuals; it has a pattern of folding similar to that of immunoglobulin constant domains, and quite extensive amino acid sequence homology with Ig constant region domains. The region of the antigen molecule away from the membrane insertion shows less similarity; this is not surprising, as it is highly polymorphic in amino acid sequence and the variation masks underlying similarities. Thus the major histocompatibility antigen molecule has, like the immunoglobulin heavy and light chains, a variable and a constant region, although the variation of the antigen is less localized than that of the antibody, and the differences occur between individuals and not between lymphocytes of the same individual.

The light chain of the major histocompatibility antigen molecule has a molecular weight similar to that of a single immunoglobulin or histocompatibility antigen heavy chain domain, a similar folding pattern, and amino acid sequence homology; it is known as β_2-microglobulin. The homologies between β_2-microglobulin and the constant region domains of the histocompatibility antigen molecule (and, to a lesser extent, perhaps also with the variable region domain) suggest the duplication and subsequent diversification of an ancestral gene for a single domain; the homologies between HLA and immunoglobulin domain structures indicate a probable common ancestry for antigen and antibody molecules. Some aspects of the immune system, at least, are probably evolutionarily very ancient and conservative. H-2 antigens of the mouse are closely similar to HLA of humans, in amino acid sequence and overall structure (Coligan *et al.*, 1981; Ploegh *et al.*, 1981; Strominger, 1980).

The genetic loci controlling histocompatibility antigen structure have been extensively analysed, and again show similarities between species. In each case, a number of relevant genes are closely linked in a *major histocompatibility complex* (MHC). The organization of the MHC regions of the mouse and human are shown in Figure 2.10 (Roitt, 1980). Of the different subregions of the MHC, H-2D and H-2K both code for conventional histocompatibility antigens of the type illustrated in Figure 2.9. Two polymorphic loci coding for histocompatibility antigens are obviously better than one, if individuality is desired; if each of the two loci has 10^3 alternative alleles within a given population, then 10^6 individuals could, in theory, be uniquely specified; and if, as is quite probable, a given individual is heterozygous for each of the loci, then four different antigens may be displayed on the surfaces of its cells. The total number of possible permutations becomes astronomical.

Duplication of histocompatibility antigen loci within the MHC is a feature of all major histocompatibility complexes so far analysed in sufficient detail. Another common feature is the assembly in the immediate vicinity of other genes relevant to immune functions: genes coding for antigens which define lymphocyte subsets and stages of differentiation, components of complement, and so on.

Between the H-2D and H-2K loci lies the I region (Uhr *et al.*, 1979). I-region gene products—the Ia, or immune-associated, antigens—form part of the array of surface components carried by, in particular, macrophages and B lymphocytes, and appear to be mediators of mutual recognition by the different cell types of the immune system. As such, they are crucial to the interactions described in the last section, and mutant I-region genes may result in defective responses to certain classes of antigen, particularly where the response to a particular antigen is T-cell-dependent.

When a macrophage presents antigen to a helper T cell, the latter must simultaneously recognize both antigen and macrophage (p. 15). Ia antigens carried by the macrophage appear to be the means of communication, and presumably interact with Ia-specific receptor sites on the lymphocyte surface. Similar recognition processes occur when the T cell in turn interacts with a B cell and induces its proliferation. Similar recogni-

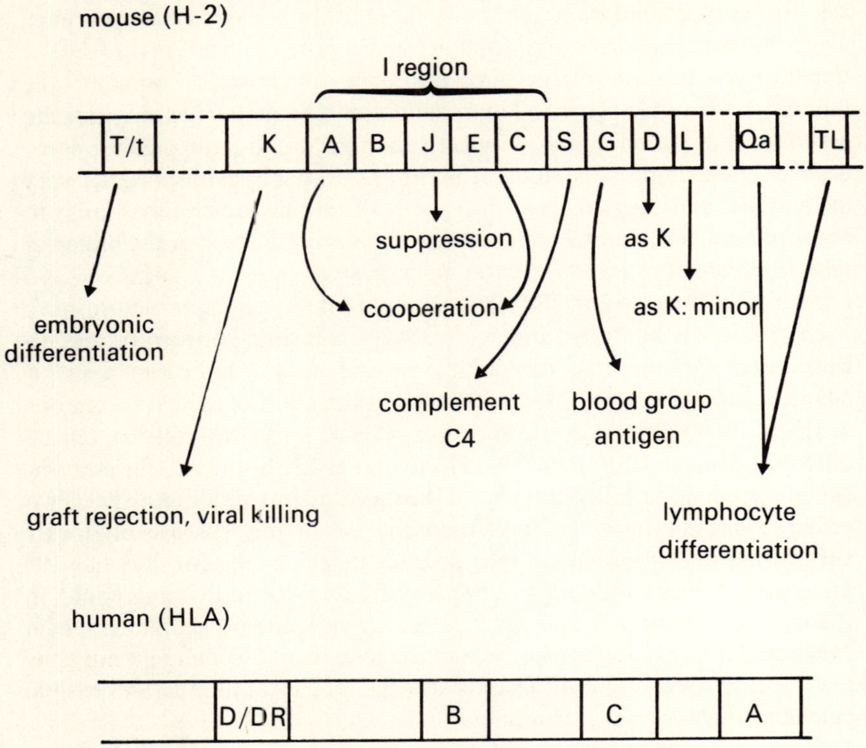

Figure 2.10 Structure of the MHC region of mouse and human. The mouse H-2 region (above) has been more fully elucidated; it lies on chromosome 17. Note that other relevant loci, although not part of the MHC itself, are linked on the same chromosome; these include the T/t locus which is involved in embryonic cell recognition processes (Artzt and Bennett, 1975; Bennett, 1975) as well as loci affecting lymphocyte differentiation. The HLA region lies on human chromosome 6. D/DR of the HLA system corresponds to mouse I region and specifies Ia type antigens, while HLA-A, -B and -C correspond to H-2D and H-2K. Human loci for complement components are known to be associated with the HLA complex, but have not been precisely localized.

tion processes occur, too, in the induction of suppressor T cells, and in their action on B cells. The component involved in helper induction appears to be a composite molecule, including one element contributed by the I-A subregion, and another controlled by both I-E and I-C subregions; the suppressor substance derives from the I-J subregion. There is also evidence that the specific soluble helper and suppressor substances which have been described above are similarly formed, possibly in further combination with molecules of the antigen to which they control the response; and that the cell surface receptors to Ia antigens are likewise, in part at least, products of the same subregions of the MHC I region (Taussig, 1980).

2.5 Immune tolerance

Exposure to antigen does not lead inevitably to rejection and a state of immunity; instead, in some circumstances, immune tolerance may be induced, with the recipient coming to accept a given antigen as if it were not foreign. Immune tolerance may persist for a considerable time, even indefinitely, and is as specific, in relation to the tolerogenic antigen, as the conventional state of immunity. An immunologically tolerant animal retains an unimpaired capacity to react against all other antigens (Roitt, 1980).

Specific immune tolerance is most readily induced in newborn animals. Injecting newborn mice of one genetic strain with cells from a mouse of a different strain can render the recipients permanently unable to reject cells or tissue grafts of the latter strain. The permanent nature of tolerance to cell injections may in some cases be maintained by the persistent presence of cells of donor origin within the host, which then constitutes a chimera, or individual composed of cells of two different origins. Injection of soluble antigen may also induce specific tolerance, but this is less likely to be permanent.

Tolerance is more difficult to achieve in adults. Generally, much higher doses of antigen are necessary ('high zone tolerance'), although repeated low doses of tolerizing antigen, perhaps in conjunction with some form of immunosuppressive treatment, may also succeed ('low zone tolerance'). The latter treatment is probably more closely analogous with the exposure of the mother to fetal antigens of paternal derivation during pregnancy.

Although the phenomenon of tolerance is well established, its mechanism is far from clear. Initially, it was suggested that if, while still immature, a lymphocyte met the antigen against which it was destined to react, it would be killed or permanently inactivated. Thus all members of 'antiself', or 'forbidden' clones would automatically be eliminated before birth, since they would inevitably encounter 'self' antigens. In an adult, with fully mature and reactive lymphocytes, a much higher dose of antigen would be required to swamp a particular clone of lymphocytes; hence the difficulty in inducing tolerance in adults.

There is some experimental evidence that very early B cells are sensitive to lower levels of stimulation than mature B cells from adults, and that after stimulation they fail to replace the surface receptor IgM molecules involved (p. 13). Lacking surface IgM they would thereafter be unable to detect the particular antigen; treatment with antigen in a very young animal would therefore have the effect of 'blinding' the members of a particular clone, rather than eliminating them.

An alternative explanation of induced tolerance of allografts is known as specific immune enhancement. This involves the production of specific humoral antibodies; these then attach to histocompatibility antigen molecules on the surface of graft cells, and so prevent recognition or access by potentially cytotoxic T lymphocytes. A precondition is that the blocking antibody should not itself be cytotoxic; it might, for instance, be antibody of a class intrinsically poor at fixing complement (such as human IgG4 or IgG1 of the mouse—Bell and Billington, 1980), or complement fixation might fail because of sparseness of the antigenic sites, since complement can only assemble itself when antigenic sites, and thus attached antibody molecules, present themselves above a critical density. Alternatively, antibody to Ia antigens might form and, by masking Ia sites on macrophages or B cells, might impede the mutual recognition processes involved in the activation of lymphocytes (p. 15, and Davies and Staines, 1976).

Blocking antibody has been implicated in a number of instances of unanticipated allograft survival, such as the persistence of tumours which carry antigens not normally found in the affected animal. The growth of some tumours is enhanced, not inhibited, by the induction of humoral antibody to the tumour-specific antigens. Antibodies which block allograft rejection might similarly help to explain fetal survival (p. 124).

The roles of different cell types in the establishment of tolerance have to some extent been elucidated. Small molecules, or antigen in a soluble form, are more likely to induce tolerance. This may be because the small molecules manage to evade the macrophages, and gain direct access to lymphocytes without being trapped, processed, and formally presented. High zone tolerance appears to reflect the unresponsiveness of B cells, while at low antigenic doses, tolerance seems primarily to be a function of T cells. This could be because of a failure of the necessary helper T cells to perform, or of the preferential induction of suppressor T cells (p. 15). As tolerance can be broken by transfer of lymphocytes from an animal which has not been rendered tolerant, the former explanation seems to apply, in some cases at least. Tolerance due to suppression of an immune response should continue in these circumstances, as the transferred cells would come under the control of the previously induced suppressor cells.

Immune tolerance may therefore be the result of several processes; whether it develops, in preference to immunity, and the basis on which it develops, will depend on the nature of the antigen, the form and amount in which it arrives, and the physiological state of the recipient at the time. As a phenomenon, tolerance is clearly of great interest in the context of survival of the fetus, considered as an allograft. This is discussed at length in later chapters. It is worth noting that, far from being an occasional failure of the immune system, tolerance is a widespread and common phenomenon: every individual discriminates its own antigens from those of others, and tolerates the former while usually rejecting the latter. Pregnancy might therefore prove to be merely a special case of the general phenomenon of self-tolerance, and might be explicable in terms of normal immunoregulatory processes.

2.6 Immunological techniques

Although many aspects of the immune system remain mysterious, there is sufficient understanding of how it works to devise practical applications, exploiting the remarkable discriminatory ability of the immune system, which have proved of great value. Although immunological techniques seem sometimes to diversify and proliferate as vigorously as lymphocytes, several methods are sufficiently important to be worth a brief description.

Frequently it is necessary to determine whether or not a given cell or tissue carries a certain antigen; for instance, whether a mammalian embryo carries histocompatibility antigens or not (p. 105). The most direct and unequivocal method of establishing this point is to prepare antibody against the antigen in question. The antibody molecules can then be chemically conjugated with a fluorescent dye such as fluorescein; this attaches to the Fc portion of the immunoglobulin molecule and so does not affect the specificity residing in the Fab regions. Labelled immunoglobulin molecules can then be allowed into contact with the questionable cells. If the appropriate antigen is in fact present, the fluorescent-labelled antibody molecules will attach and their presence and location is revealed under ultraviolet light by virtue of the fluorescence of the labelling dye.

A more sophisticated version of this immunofluorescence technique is known as *indirect immunofluorescence* (IIF). Here unlabelled specific immunoglobulin is allowed to attach to the antigen (if the latter is present); the attachment of immunoglobulin—hence the presence of antigen—is made manifest by adding fluorescent-labelled anti-immunoglobulin antibody. (Immunoglobulin molecules, being protein and highly varied, can of course act also as antigens.) The advantage of IIF is that it gives more intense fluorescence, and is therefore particularly valuable when dealing with small concentrations of antigen. It is also convenient, in that only one immunoglobulin needs to be conjugated with fluorescein, and can then be used (indirectly) for locating a range of antigens, for which unlabelled antibody is available (Roitt, 1980).

A handicap of immunofluorescence techniques is that in practice when antigen is injected into an experimental animal, a range of different antibody molecules, of different degrees of specificity and affinity, is usually the result. Better definition of the antibody being used would be desirable. Recently it has become possible to refine this technique by inducing antibody of a single type, the product of the progeny of a single B cell.

The initial difficulty in achieving this stemmed from the fact that a B lymphocyte in culture has a limited life. This made it impossible to breed up the offspring of a single progenitor cell and produce large amounts of antibody from the cells of a single clone. Most normal cells, B cells included, do not make good laboratory animals; the only fully domesticated cell types belong to tumour cell lines, many of which have been established in laboratory culture for many years. Fusion of a normal B cell with a cell of a long-established B-cell tumour line can result in a hybridoma cell line: the offspring of the unnatural union combine the antibody specificity of their normal parent with the immortality of the tumorous parent. Unlimited quantities of identical antibody molecules can now be obtained. The value of monoclonal antibody availability is potentially immense, and many fields of biology have already felt the impact.

Another question which may be asked, particularly in the investigation of the immunology of reproduction, is whether or not an animal has responded to a certain antigen with the production of antibody. The presence of a specified antibody can be established by a variety of methods. If the antigen is soluble, then combination with antibody may cause the precipitation of insoluble antigen–antibody complex. A common method is to allow antigen and (putative) antibody to meet by diffusion through agar. If more than one antigen–antibody combination is present, and if (as is likely) different antigens diffuse through agar at different rates, then more than one zone of precipitation will appear.

A particulate antigen in suspension will follow an analogous process and agglutinate in the presence of relevant antibody. This is a useful and rapid method of establishing the presence of antibody against spermatozoal antigens (p. 30). Soluble antibody against cell surface antigens can be detected, after combination with its target, by using fluorescent-labelled anti-immunoglobulin antibody, or by assessing the fixation of complement to the attached antibody. In the latter case, fluorescent-labelled antibody against complement is valuable.

The cellular immune response is less straightforward. Intrinsic capacity to mount a cellular immune response can be assessed by treating lymphocytes with non-specific, or polyclonal, activators, such as the plant-derived mitogens phytohaemagglutinin (PHA) or concanavalin A (Con A). The degree of cellular response to specific antigen, rather than to mitogen, can be assessed by the *mixed lymphocyte* (or *leucocyte*) *reaction* (MLR). This involves extracting leucocytes from two genetically dissimilar animals and mixing them in culture. Each set of leucocytes responds to antigens carried by the other; a reciprocal stimulation of proliferation follows, and can be measured either in terms of the rate of uptake of radioactively-labelled DNA precursors, or by assaying the concomitant release of lymphokines. Lymphokines are substances produced by reacting lymphocytes, which have the ability to summon macrophages and other cells to their aid. One in particular, macrophage *migration inhibition factor* (MIF) acts to detain passing macrophages in the immediate vicinity of the responding lymphocytes. MIF can be accurately assayed using the movement of guinea pig macrophages in closely defined circumstances, and provides the basis for a convenient bioassay of the intensity of a mixed lymphocyte reaction (Cohen, 1979; Roitt, 1980).

The responding cells in the MLR are predominantly T lymphocytes and the antigens to which they respond are Ia antigens (p. 19) carried mainly by B cells and macrophages. If, in addition, the two donor animals differ with respect to H-2D or H-2K antigens, or their equivalents in other species (see p. 19), cytolysis may also occur, and can be used as the basis of a form of measurement of the intensity of the response. This development of the basic MLR approach is known as *cell-mediated lympholysis* (CML) and has been used as a method of gauging histocompatibility between putative graft donors and recipients (Roitt, 1980).

As described, the MLR would be a mutual one, with lymphocytes of each donor type reacting to antigens of the other. If the leucocytes from one donor are prevented from responding before mixing—for example, by being inactivated by irradiation—the resulting one-way MLR gives information about the reaction of cells only of the other donor. The timing of proliferation can also be exploited to yield further information. If significant proliferation of the responding cells takes place within 24 or 48 hours of mixing, it can be inferred that the responding cells have previously been sensitized to the stimulating antigen. Proliferation which reaches a significant extent only 5 days or so after mixing, in contrast, suggests that sensitization has occurred *in vitro* to antigens that the responding cells have not encountered before. This difference in timing has been used to distinguish actual maternal sensitization to fetal antigens, as opposed to the *potential* for sensitization (e.g. Birkeland and Kristoffersen, 1980*b*; Wegmann *et al*., 1979; and see p. 124).

CHAPTER THREE

IMMUNOLOGY AND REPRODUCTION IN MALE MAMMALS

3.1 Structure of the male reproductive system

The male reproductive system of a representative mammal is shown diagrammatically in Figure 3.1. The bulk of each testis consists of a convoluted mass of fine tubules of *seminiferous epithelium*, within which spermatozoa are produced. These seminiferous tubules are numerous—approximately 800 in each human testis—and this, together with their

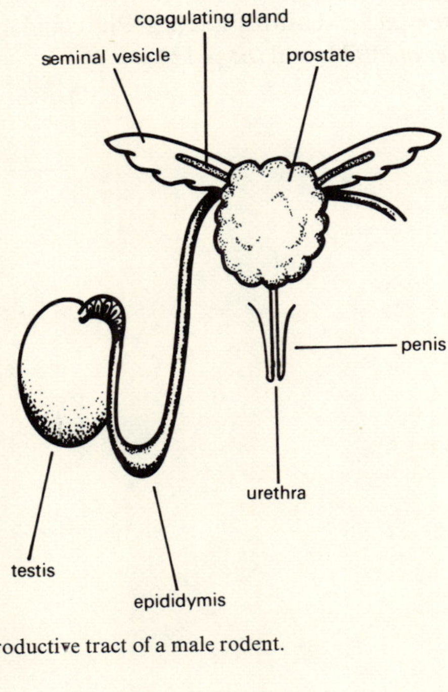

Figure 3.1 The reproductive tract of a male rodent.

length, provides a considerable surface area over which spermatogenesis can proceed. The rate of sperm production in the human testis has been estimated at around 50 000 a minute. The large numbers in which sperm are generally found is a point of some immunological importance.

Cells other than germ cells also occur in the testis. Between the seminiferous tubules lies the major endocrine tissue, the *Leydig* (or interstitial) *cells*. These produce steroid hormones, mainly androgens. Blood vessels and lymphatics ramify among the interstitial cells, but do not penetrate the seminiferous tubules; nutrients are supplied to the developing germ cells only indirectly.

A further category of somatic cells are the *Sertoli cells*, which lie within the seminiferous tubules, attached to the basement membrane. Figure 3.2 shows diagrammatically the arrangement of Sertoli cells within the tubules, and their relationship with each other. Tight junctions between adjacent Sertoli cells divide the interior of the tubule into two concentric compartments. The outer, basal, compartment is separated by the tight junction zone from an inner compartment which includes the lumen proper, as well as the upper levels of each Sertoli cell. Movement of substances between these two compartments by free diffusion between the cells is negligible; thus by their secretory activity the Sertoli cells effectively control the composition of the fluid which enters the seminiferous tubules. In many respects the fluid in the lumen differs from both blood serum and from the tissue fluid immediately outside the tubules and in the space

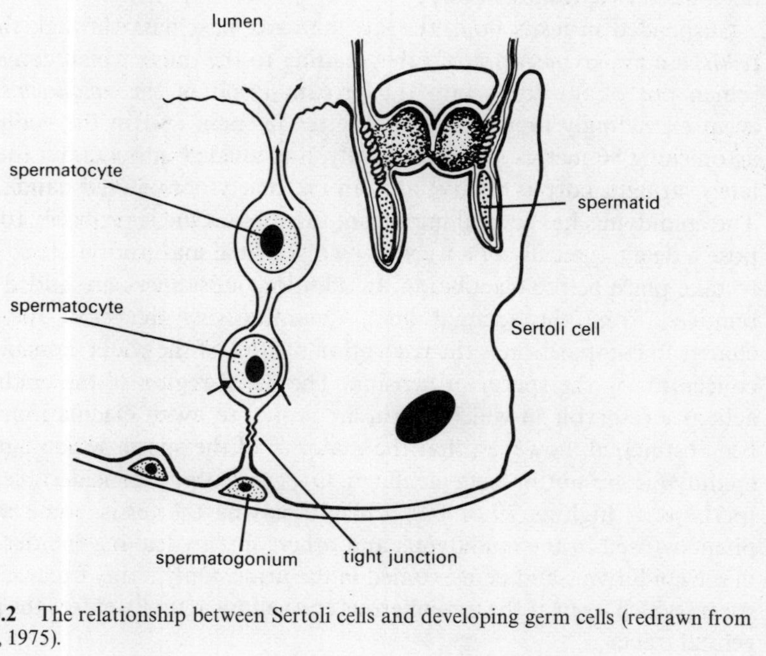

Figure 3.2 The relationship between Sertoli cells and developing germ cells (redrawn from Fawcett, 1975).

between the basement membrane and the level of the tight junction zone; for example, the albumin concentration of the tubule fluid is only some 1–1.5% that of blood serum. The barrier at the level of the tight junction maintains a microenvironment which is essential for the development of viable spermatozoa.

The precursors of spermatozoa originate in the basal compartment. Here, as spermatogonia, they enter upon several proliferative cycles of mitosis; different spermatogonial stages are designated Type A, intermediate, and Type B spermatogonia. Older spermatogonia move progressively out from the basal membrane, towards the centre of the tubule. Tight junctions form behind them and the older tight junctions relax ahead; in this way the germ cells pass from one compartment to the other without compromising the integrity of the barrier. On entry into the adluminal compartment, spermatogonia become primary spermatocytes and enter meiosis. Still tending to move towards the lumen, they complete meiosis as secondary spermatocytes, then, without further division, become spermatids. Early spermatids, although haploid, look like fairly typical cells. Soon, however, they begin to develop tails, and acrosomes—the latter being cone-like structures which contain enzymes necessary for penetrating the cumulus cells and zona pellucida (p. 51) which surround the egg. Spermatids thus become mature spermatozoa (Figure 3.3), which are then released from their enveloping Sertoli cells into the tubule lumen, in the process leaving most of their remaining cytoplasm behind within the Sertoli cell as a 'residual body'.

Suspended in testis fluid, the spermatozoa next pass through the *rete testis*, a complex network of tubes leading to the ducts which convey the semen out of the testis into the proximal end of the *epididymis*. This is an exceedingly long tubule—8 metres in man, and in the stallion an astonishing 86 metres. Conventionally, it is divided into a caput (head), a fairly straight corpus (body) and an extremely convoluted cauda (tail). The epididymis has several important functions. One is probably to interpose a delay, typically of a week or two, for final maturation of the sperm to take place before ejaculation. In addition, substances are added to, or removed from, the seminal fluid. Quantitatively the most important change in composition is the resorption of 90% of the water, considerably concentrating the sperm suspension. The distal region of the epididymis acts as a reservoir in which sperm are stored to await ejaculation. It has been estimated, however, that the majority of the sperm which enter the epididymis are not in fact ejaculated. Over and above considerable losses (perhaps as high as 70 or 80%) before leaving the testis, some may be phagocytosed in the epididymis and others lost by leaving the distal end of the epididymis and being voided in the urine. Only a tiny fraction of the sperm which form at the seminiferous epithelium actually enters the female genital tract.

During coitus, sperm are propelled out of the cauda epididymis through the vas deferens to the urethra and thereby to the exterior. As this takes place, the bulk of the seminal plasma (95% or more by volume) is added by the accessory organs, notably the seminal vesicles and prostate in man. Other species have variations on the accessory gland theme, such as the coagulating glands of small rodents (Figure 3.1). Additions to the seminal fluid at this stage include a cocktail of nutrients such as fructose and citric acid, enzymes such as those involved with the clotting and subsequent

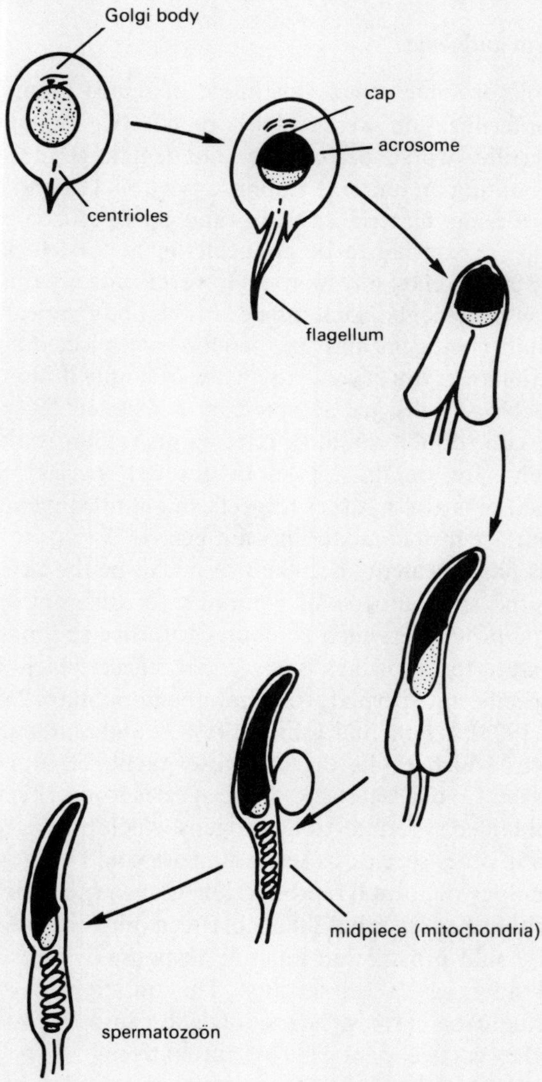

Figure 3.3 Sperm maturation (spermiogenesis) in the rat (redrawn from Monesi, 1972).

liquefaction of semen in the female tract, prostaglandins, and other substances which differ from species to species.

What is deposited in the female, then, is a dense suspension of cells ($50–60 \times 10^6$ in a mouse ejaculate of 0.1 ml) in a medium of considerable complexity. The spermatozoa themselves, and many of the components of the seminal plasma, might be expected to be antigenic and therefore capable of inducing an immune response which, potentially, could impede reproduction (Austin and Short, 1972*a*; Hogarth, 1978).

3.2 Are sperm antigenic?

If spermatozoa, or some other constituent of semen essential for sperm survival or for fertilization, were to evoke an effective immune response in the female, fertility would be reduced. The female genital tract is fully capable of mounting an immune response (section 4.2) and of destroying antigenically foreign material. Semen, and spermatozoa in particular, might for some reason fail to be antigenic, or at least to be sufficiently strongly antigenic to elicit a response. However unlikely this might seem, there are a few precedents; some tissues in the body appear not to carry histocompatibility antigens and are therefore not rejected after allotransplantation. Human erythrocytes are a case in point: if blood groups are matched, red blood cells are accepted by a recipient even when other tissues of the same donor would be rejected (as is almost always the case in humans who are, on the whole, an outbred species). Spermatozoa, unconventional cells also in other respects, might unconventionally fail to express any surface histocompatibility antigen.

In fact, this has repeatedly been shown not to be the case. Injection of rabbits with the spermatozoa of a number of different species causes antibody to be produced which combines with live spermatozoa *in vitro* and agglutinates them or has some other effect which indicates the presence of specific antispermatozoal immunoglobulin (p. 24)—see Cohen and Hendry (1978), Edidin and Johnson (1977), and Shulman (1972).

The antigens which evoke this response might, of course, merely be xenoantigens (p. 13) characteristic of the species from which the spermatozoa were obtained. Spermatozoal antigens which produced an immune response only in other species, as xenoantigens, would be largely irrelevant to the immunology of normal reproduction. Only a substance which is *not* ubiquitous within a species, but differs between different individuals of the same species, could provoke an immune response as a result of mating, which could adversely affect fertility. The question is not, therefore, whether spermatozoa carry substances which can prove to be antigenic, but whether they carry alloantigens (or autoantigens).

Further investigation reveals a situation of considerable complexity. Human spermatozoa carry a mixed cargo which includes antigens of the major blood group and histocompatibility systems. The presence of antigens of the ABO blood group system is shown by the agglutination of mobile and actively swimming sperm in the presence of certain prepared antisera against appropriate blood group antigens (Edwards *et al.*, 1964; Shulman, 1972). The ABO blood group system is peculiar in that antibodies against the major determinants are naturally-occurring without being induced by the presence of antigen. People of blood group A, for instance, carry antigen A on their erythrocytes and have anti-B antibody in their blood serum. For some reason, these naturally-occurring antibodies fail to agglutinate spermatozoa which carry the relevant A or B antigen. This is perhaps fortunate, since blood group antibodies are secreted in the cervical mucus, through which spermatozoa must pass to enter the uterus (p. 55)—see Parish *et al.* (1976).

Controversy has surrounded the origin of the blood group antigens of spermatozoa. The antigens could have been produced by the spermatozoa themselves, or by their precursors; or they could have been acquired by adsorption to the sperm surface from the surrounding seminal plasma. Recent evidence suggests that both views may be correct. About 76% of the human population are 'secretors' with respect to blood group antigens: that is, significant amounts of antigen are found in solution in all mucous secretions, including seminal plasma. Washed spermatozoa from a 'nonsecretor' of blood group O, when incubated in seminal plasma of a 'secretor' of group A, acquire group A antigenicity from the seminal plasma, while (according to some accounts at least) the spermatozoa of group A or group B nonsecretors completely lack blood group antigens. The obvious inference is that spermatozoa which carry blood group antigens have passively acquired them from the soluble antigenic material secreted into the seminal plasma (Boettcher, 1968; Edwards *et al.*, 1964).

Other workers, using in many cases different techniques (which makes direct comparison difficult) have presented evidence that antigens of the ABO system are in fact intrinsic components of the spermatozoal membrane, being present (albeit in small amounts) on the spermatozoa of nonsecretors. It has also been found that, within the ejaculate of a man of heterozygous blood group genotype (group AB) the spermatozoa may be heterogeneous; they carry predominantly either A antigen, or B antigen, but not equal amounts of both. This is what would be expected if the blood group phenotype of a spermatozoon was determined by post-meiotic expression of its genotype: each haploid sperm could express only one of the two alternatives of the heterozygous diploid genotype of the man in question. Post-meiotic RNA synthesis does occur at much the same level (relative to the amount of DNA present per cell) as in premeiotic diploid cells, so haploid expression appears feasible. However, it would be surprising if the spermatozoa fell into two sharply distinguishable categories on this basis, since spermatocytes and spermatids do not physically separate after meiosis but retain continuity of their surface membranes, and cytoplasm, through persistent cytoplasmic bridges. Most of the evidence for haploid expression can be explained equally well in terms of quantitative variation in antigenic expression. The question remains controversial (Shulman, 1972).

Antigens of other blood group systems have also been detected on human spermatozoa, among them those of the Lewis, MNSs and P systems. Antigens M, N and Tja (of the P system) are absent from the seminal plasma even of secretors, so are presumably produced by the spermatozoa themselves. One conspicuous omission appears to be the antigen D of the Rhesus blood-group system; a pity, since if this antigen was expressed in a haploid way on spermatozoa, it might eventually have proved possible to use it as a means of identifying, and removing from an ejaculate, those sperm which would cause a Rh-positive fetus to be conceived by a Rh-negative woman. Rhesus haemolytic disease of the newborn (see Chapter 6) might in this way be avoided.

The position in other species has been less thoroughly investigated. Blood group antigens have been located on the spermatozoa of some breeds of cattle (but not others). Rabbits appear not to have antigens common to both red blood cells and spermatozoa, although it has been suggested that this could be due to the lack, in the particular strains used, of the equivalent of 'secretor' status of humans (Padma, 1972).

Injection of mice of one genetic strain with a suspension of spermatozoa collected from the genital tracts of males of a different strain accelerates the rejection of subsequent skin allografts from mice of the same strain as the sperm donors. In one series of experiments of this sort, for instance, the median rejection time of allografts to mice which had each received a prior injection of 10^8 spermatozoa was 12 days, compared with 21 days for the control (uninjected) animals (Beer and Billingham, 1976). Successful presensitization to skin allografts demonstrates that spermatozoa and skin carry antigens in common, and that these include the histocompatibility antigens, responsible for eliciting graft rejection.

More specifically, it has been shown that mouse spermatozoa carry various antigens including the major histocompatibility antigens of the H-2 system (H-2D and H-2K) and, in addition, antigens specified by minor loci such as H-3 and H-13 (Herr and Eddy, 1980; Vojtíšková and Pokorná, 1972). As histocompatibility antigens are generally attached to cell membranes, and are thought not to occur in significant quantities in solution in seminal plasma, their presence on spermatozoa is presumably due to synthesis by the germ cells themselves. (However, as recent research has established that MHC antigens can be shed from the surfaces of various cells (Emerson *et al.*, 1980), perhaps this should not be taken for granted.) Mouse spermatozoa also carry antigens of the Ia type (Fellous *et al.*, 1976), and possibly also an antigen common to spermatozoa, early embryos, and tumour cells, which may be a gene product of the T/t locus (Gachelin *et al.*, 1976; Goodfellow *et al.*, 1979; and see pp. 40 and 109).

Spermatozoa used in the presensitization experiments to establish the presence of spermatozoal H-2 antigens were obtained from the cauda epididymis and vas deferens. Sperm collected in this way are contaminated with about 0.1% of non-spermatozoal cells, of which 10% are leucocytes of various types. It has been argued that it is these cells, rather than the spermatozoa themselves, which were antigenic, with the sperm contributing only by potentiating the response to passenger leucocytes (Beer and Billingham, 1974).

Immunofluorescence has, however, disproved this hypothesis, and directly revealed the presence of MHC antigen on mouse spermatozoa. Comparison of the ability to absorb cytotoxic anti-H-2 antibody from solution has made it possible to estimate the number of antigenic sites on spermatozoa, compared with the number carried by other cell types. Mouse sperm carry about one tenth as much H-2 antigen as lymphocytes separated from regional lymph nodes. As complement-mediated destruction depends on the density of sites to which antibody can attach, this might militate in favour of sperm survival in the female genital tract; on the other hand, as sperm have a smaller surface area than lymphocytes, lower total antigenicity with respect to H-2 antigens does not necessarily imply lower density of antigen.

Human spermatozoa have similarly been shown, by immunofluorescence and related techniques to carry HLA antigens (Kerék *et al.*, 1973). These are distributed patchily over the head and middle piece of the sperm cells, in contrast to 'species' antigens which spread over both head and tail. Although there have been reports of soluble HLA material in human seminal plasma, it seems probable that in humans, too, the MHC antigens of spermatozoa are endogenous. Also as in mice, the spermatozoa in a single individual's ejaculate may be heterogeneous with respect to the HLA type carried, which is an indication of post-meiotic expression of the haploid germ cell genome. I-region antigens, in this case HLA-D (see p. 20) are also carried on human sperm (Fellous and Dausset, 1970; Levis *et al.*, 1976).

Skin from male mice may be rejected by females of the same genetic strain, where grafts between males, between females, or from female to male are readily and permanently accepted. This is due to a transplantation antigen associated with possession of a Y chromosome, known as H-Y antigen (Austin and Short, 1972*b*; Hogarth, 1978). (Male mammals have the sex chromosome constitution XY, females XX.) H-Y antigen is thought to have a crucial role in the primary determination of sex, but as a histocompatibility antigen its significance is normally trivial. Only when exposed against the background of genetic uniformity of a highly inbred strain can discrepancy at the H-Y locus alone cause graft rejection.

Prior sperm injection can lead to accelerated rejection of a male-to-female skin graft, and specific antisera have been used to reveal the distribution of H-Y sites on spermatozoa. In this case, the location of attached antibody against H-Y was shown by tagging it with particles of tobacco mosaic virus (TMV), whose characteristic shape makes them readily recognizable under the electron microscope. H-Y antigen in mice is almost entirely restricted to the acrosomal cap. It is present in varying amounts on the sperm within an ejaculate, and about 20 to 40% have virtually none. As only half of the sperm within an ejaculate carry a Y chromosome, this again suggests postmeiotic expression of the genome; as well as a potential method of predetermining the sex of offspring by using spermatozoal H-Y antigen as a basis for separating sperm into male- and female-producing (Koo *et al.*, 1973).

In addition to antigens such as those of the blood group and histocompatibility antigen systems, which spermatozoa hold in common with somatic cells, there is also a class of sperm-specific antigens, peculiar to spermatozoa or their immediate germ cell precursors. These are autoantigenic, that is they are potentially capable of inducing an immune response in the body of the male which produced them; outside the testis, they count as 'foreign' to the rest of the body. An autoimmune response against these spermatozoal antigens is normally prevented by their sequestration behind a 'blood-testis barrier', represented chiefly by the tight junction zone of the Sertoli cells (p. 27). Only rarely is this barrier breached; bypassing it artificially by injecting spermatozoa outside the confines of the testis, and in consequence evoking an immune response, is one of the principal criteria which establishes the existence of spermatozoal autoantigens.

Using as their experimental animals guinea pigs, which are very susceptible to the induction of an anti-spermatozoal autoimmune reaction, Voisin and his associates have identified four autoantigens of spermatozoa. These they designated P, S, T and Z. Of the four, Z has proved particularly elusive, and tends to disappear during the complicated extraction procedures used, so although its existence has been established, little is known of it except that it may be a glycoprotein. Of the other three, P is a protein of molecular weight probably less than 60 000, S a sugar-rich glycoprotein of high molecular weight, and T possibly a lipoprotein (Johnson, 1975*a*; Voisin *et al.*, 1974).

Antibodies against T are spermagglutinating, which indicates that T is carried on the outside of the spermatozoa: otherwise it would be impossible to agglutinate live and intact sperm. This inference is confirmed by immunofluorescence. Anti-T autoantibody is also capable of fixing complement and so triggering complement-mediated lesions in the surface membranes of the spermatozoa, especially in the acrosome region of the sperm head. These prove lethal. The progress of destruction is remarkably rapid; some 20% of the sperm develop lesions within 30 seconds, 40% within one minute, and 80% within 5 minutes of the addition of antibody and complement.

Autoantibody against P antigen is also complement-fixing, as is a small proportion of the autoantibody against S. However, as both P and S lie below the surface of the sperm, within the acrosome, these autoantibodies can neither agglutinate intact sperm nor initiate an attack on them. They may contribute to later stages of the destruction, when autoantibody against T has already breached the outer membrane of the sperm and exposed internal antigens to attack (Le Bouteillier *et al.*, 1975; Russo and Metz, 1974).

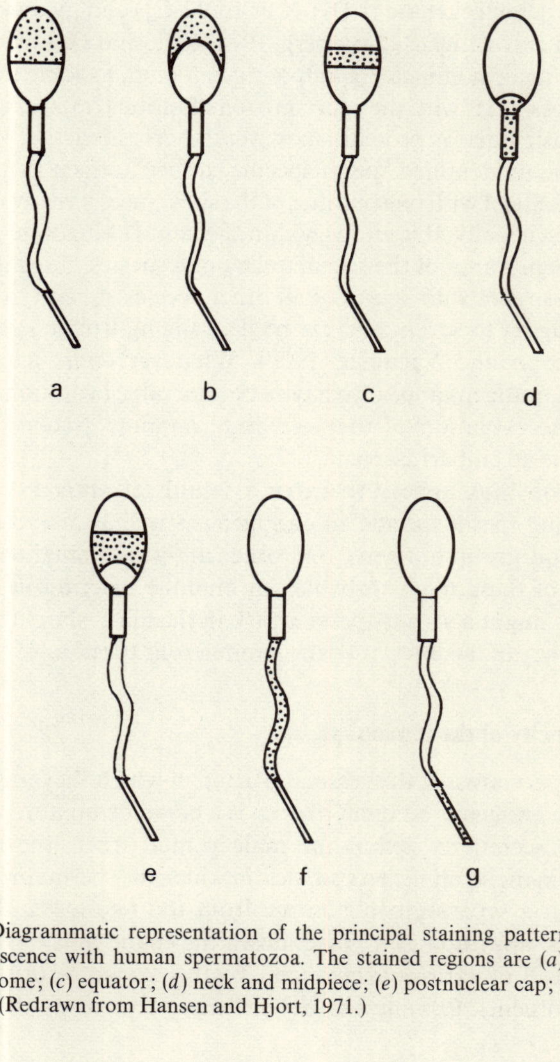

Figure 3.4 Diagrammatic representation of the principal staining patterns obtained by immunofluorescence with human spermatozoa. The stained regions are (*a*) acrosome; (*b*) anterior acrosome; (*c*) equator; (*d*) neck and midpiece; (*e*) postnuclear cap; (*f*) tail; and (*g*) tail end piece. (Redrawn from Hansen and Hjort, 1971.)

A range of other autoantigens of guinea pigs have been identified, of which some are known to be glycoproteins, and may be related to one or other of the antigens described by Voisin (Hagopian *et al.*, 1975, 1976; Jackson *et al.*, 1975).

Spermatozoal autoantigens in other species have generally been less thoroughly studied, but their presence has been shown in all species investigated. In man, for instance, two highly basic nuclear proteins, protamines 1 and 2, have been identified as autoantigens (Kolk and Samuel, 1975). Other autoantigens have not been chemically characterized, but have been shown by immunofluorescence to be fairly distinctively distributed within the spermatozoon (Figure 3.4) (Hansen and Hjort, 1971; Myles *et al.*, 1981). In addition, enzymes such as the sperm-specific form of lactic dehydrogenase (LDH-X or LDH-C_4) and hyaluronidase are also potential autoantigens (Goldberg, 1974; Kille and Goldberg, 1979).

Antibodies against guinea pig autoantigen S (and, to some extent, those against T) cross-react with the spermatozoa of rabbits, rats, mice and men, and sperm basic nuclear proteins show similar cross-reactivity (Pruslin *et al.*, 1979). One unidentified sperm-specific surface antigen (which may or may not be identical with one or other of the substances already mentioned) appears to be virtually ubiquitous within the animal kingdom. Antisera to the plasma membranes of the spermatozoa of a species of sea urchin were found to cross-react with sperm of all other species studied: a total of 28 species, belonging to seven different phyla, ranging from coelenterates to chordates (Lopo and Vacquier, 1980). Whatever their function, some sperm-specific antigens appear to have a considerable evolutionary history. Only with the evolution of the vertebrate immune system might such antigens prove an embarrassment.

Spermatozoa thus appear to carry a wealth of antigens of different types, including sperm-specific autoantigens as well as histocompatibility antigens, blood group antigens, and other antigens shared with somatic tissues. Any of these might provoke an immune reaction in the female. Autoantigens might also put sperm at risk in the male, should they escape from the privacy of the interior of the seminiferous tubules.

3.3 Antigenicity of the seminal plasma

Besides the spermatozoa, the seminal plasma in which they are suspended might also be antigenic. Seminal plasma is a complex mixture, originating from several secretions within the male genital tract: the initial fluid secreted within the seminiferous tubules, in which the spermatozoa embark, is mixed with a separate contribution from the rete testis, then further modified in composition as it passes down the epididymis by the addition and withdrawal of various substances. To this the accessory glands add further constituents, forming the bulk of the ejaculate, as the sperm are propelled out during coitus. The major accessory glands in man are the prostate and the seminal vesicles, with the bulbo-urethral and other glands making relatively minor contributions. In other species, the relative importance of the different components may be different; rats and mice have prominent coagulating glands, while the dog has a conspicuous prostate but no seminal vesicles.

Many of the substances secreted are enzymes or other large molecules, and these might well be antigenic. Various methods have been employed to study the number and nature of seminal plasma antigens of accessory gland origin. If human seminal plasma is injected into rabbits and the resulting antiserum used to separate seminal constituents by immunoelectrophoresis (a combination of the gel diffusion method described in section 2.6, p. 24), with further separation of protein components according to their electrophoretic mobility, twelve xenoantigenic components can be distinguished.

To eliminate the possibility of the inclusion of antigens actually derived from spermatozoa, similar experiments were carried out using rabbit antisera against the seminal plasma of men whose semen was completely devoid of spermatozoa. In this azoospermic seminal plasma 13 to 15 components were distinguished. Some of these occurred elsewhere in the body as well. When the rabbit anti-human seminal plasma antisera were absorbed with human serum and liver extracts, the number of seminal plasma components which still reacted fell to six. Absorption removes from the mixed antisera those antibodies raised against elements of the seminal plasma which react also with antigens present in blood serum or liver. Assuming these to be antigenically representative of the rest of the body, the six remaining components of seminal plasma represent antigens which do not occur elsewhere, and which might in consequence act as autoantigens (Isojima *et al.*, 1974).

If the antiserum containing antibodies against the six presumed seminal-plasma-specific antigens was allowed to diffuse in agar against human milk protein, two precipitin bands appeared, one strong and the other weak, while further absorption of the antiserum with human milk reduced the apparent number of seminal plasma antigens from six to four. Seminal plasma must therefore have two antigenic components in common with milk. One at least has been firmly identified as lactoferrin (Hekman and Rümke, 1969; Tauber *et al.*, 1975). This is an iron-containing β-globulin protein which is probably important in the transmission of essential iron to the suckling baby; it also has some bacteriostatic properties which may be useful to the infant. Seminal lactoferrin might share these properties, and help to control bacteria which enter the female genital tract during copulation. Although there are slight differences between lactoferrin of milk and its seminal plasma homologue, the two are antigenically indistinguishable.

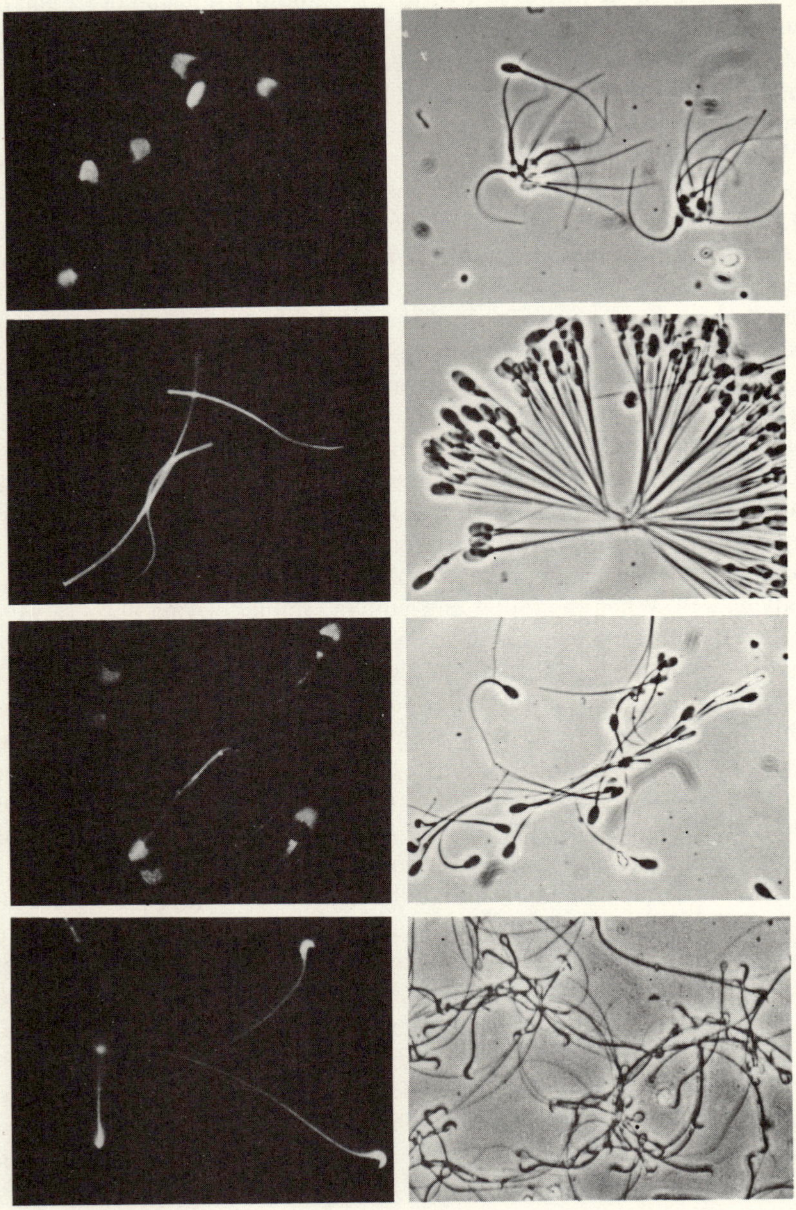

Figure 3.5 The different immunofluorescence staining patterns, and different forms of agglutination produced by different antisera against rabbit spermatozoa (from Moore, 1981, with permission).

Lactoferrin is absent from the semen of men who lack seminal vesicles, so is presumably added to semen by these organs. It is present also on the surface of ejaculated spermatozoa where, unlike the spermatozoal antigens already described, it is distributed generally over the entire surface rather than being restricted to certain zones (cf. Figures 3.4 and 3.5). It is not found on sperm collected directly from the testis, or in the epididymis. The obvious inference is that it is secreted into the seminal plasma by the seminal vesicles, and adsorbed from the seminal plasma onto the sperm during ejaculation. For this reason seminal lactoferrin is referred to as a sperm-coating antigen (SCA), or sometimes as scaferrin, in acknowledgement of its slight differences from milk lactoferrin (Hekman and Rümke, 1969; Roberts and Boettcher, 1969).

Specific antigenic contributions of other accessory glands are less well known, although there is evidence that the prostate adds at least three potential autoantigens to human seminal plasma (Ablin *et al.*, 1970; Li and Behrman, 1970; Shulman and Bronson, 1969).

In the rabbit, glycoprotein secretions of the distal caput and corpus epididymis also coat sperm. Three distinguishable components appear, and each coats the spermatozoa in a distinctive way. One, designated R1, coats only the anterior acrosome region; antibody against R1 causes head-to-head agglutination of spermatozoa. R2 is found on the anterior acrosome and midpiece, and to a slight extent on the tail; anti-R2 antibody, as might be expected, causes a mixed agglutination. R3 is restricted to the tail and is involved in exclusively tail-to-tail agglutination (Figure 3.5)—see Moore (1981).

Rabbit seminal plasma autoantigens derived from the accessory organs have been identified, including contributions from the seminal vesicles (which contain 2000 times the seminal plasma concentration of one antigen) and the prostate (Riera *et al.*, 1967; Shulman *et al.*, 1966).

Semen is therefore a complex mixture of antigens from different sources. Some of these are autoantigenic and could therefore in certain circumstances provoke an immune response in the male which produced them. Equally, they could evoke a deleterious immune response by the female; as, indeed, could components of the semen which are not autoantigenic but alloantigenic, such as the histocompatibility antigens or blood group antigens carried by spermatozoa. The possibility of autoantigenic responses to seminal antigens is discussed in the following section, and the female's response to semen in section 4.4.

3.4 Autoimmunity in the male

Any of the antigens carried by spermatozoa, or contained within seminal plasma, might induce an immune response in the female and impede fertility. Only the autoantigens are potentially capable of provoking such a response in the male. In experiments *in vitro*, testicular cells of the mouse appear to have a propensity for inducing a non-specific form of immune tolerance in competent autologous lymphocytes which confront them. The precise antigens involved are not known, but it is an intriguing speculation that they may include a group of antigens common to spermatozoa and embryonic tissues (Gachelin *et al.*, 1976), since certain embryonic cells show similar immunosuppressive capacity *in vitro* (Dooher *et al.*, 1981; Hurtenbach *et al.*, 1980).

It is, as always, difficult to ascertain the relevance of experiments *in vitro* to the behaviour of cells in their normal context. It seems likely that autoimmune responses to testis antigens normally fail to occur largely because of lack of access of antigen to the immune system. Such access is generally prevented by the blood–testis barrier, the anatomical basis for which is the cordon of tight junctions between adjacent Sertoli cells (sections 3.1 and 3.2). Injection of male guinea pigs with fluorescent-labelled albumin has shown that protein fails to penetrate the barrier. Similar tight junctions occur in the epithelia lining the downstream parts of the male tract, such as the rete testis and the vas deferens. The efferent barrier is, however, probably less effective here than in the testis, as labelled proteins do in fact penetrate the rete and may reflux to the seminiferous tubules (Friend and Gilula, 1972).

However, from the point of view of spermatozoal autoantigens escaping from the ducts of the genital tract and inducing an immune response, the position may not be quite so simple. Although lymphocytes are completely absent from normal seminiferous epithelia, there does appear, at least in rats, to be a resident population of lymphocytes within the epithelia lining more distal stretches of the male genital tract such as the rete testis and epididymis (Dym and Romrell, 1975). These lymphocytes are very much more common in sexually mature males than in animals which have not yet produced spermatozoa and associated antigens. It is possible that they supplement a relatively ineffectual tight-junction barrier and help to segregate spermatozoal autoantigens from the rest of the body.

Antibody is found in human semen, including IgG, IgA and IgE, but excluding the larger molecules of IgM. The antibody concentration of epididymal fluid appears very low, while the seminal immunoglobulin content of vasectomized men is normal; these facts together suggest that seminal immunoglobulin enters the male tract below the vas deferens. The prostate is probably the main point of entry. In guinea pigs, immunofluorescent techniques have shown a similar situation, with immunoglobulin absent from the lumen of the seminiferous tubules, rete testis or epididymis, and present only in the secretions of the coagulating gland, prostate and seminal vesicles (Johnson, 1972; Tauber *et al.*, 1975).

Even in seminal plasma, antibody concentrations are not high compared with blood serum. In men, IgG is generally around 1–2% of serum levels and IgE at about 30%; the high level relative to IgG suggests that the entry of IgE into seminal fluid is actively assisted. IgA is present in relatively large amounts—up to several times the serum concentration—and probably derives from local synthesis and secretion rather than diffusion, assisted or otherwise, from the blood (Hjort, 1977).

The blood-testis barrier appears therefore to act by preventing egress of autoantigenic material from the seminiferous tubules. Distal to the testis, a partially effective afferent barrier is perhaps supplemented by epithelial lymphocytes. The barrier between the contents of the male genital tract and the rest of the body also restricts the efferent passage of immunoglobulins, if they form despite these precautions, to the lower reaches of the tract, through which spermatozoa pass only briefly.

The existence of an effective barrier is a necessary condition for the prevention of an autoimmune response. A further condition is that the autoantigens lie only on the luminal side of the barrier. This has been demonstrated in the guinea pig by using a fluorescent-labelled antibody against Voisin's autoantigen P (p. 34). Autoantigen P occurs only within the lumen of seminiferous tubules, epididymis and vas deferens, and neither within the epithelial cells of these ducts nor outside the ducts (Johnson, 1972).

A further requirement is that no autoantigen should be produced by the germ cells until after the blood–testis barrier has been established. The presence of the barrier is necessary to preserve the specialized micro-environment required for the maturation of germ cells. It is therefore fully established by the time of puberty, before the first appearance of spermatocytes, spermatids, or spermatozoa. Several studies have confirmed the appearance of spermatozoal antigens only after the developing testis has reached the stage of producing spermatocytes for the first time. Guinea pig spermatozoal autoantigens S, P and T appear, for example, only from the 26th day after birth, coinciding with the debut of late spermatocytes (Radu and Voisin, 1975). Much the same has been shown to be true of a variety of other autoantigens in rabbits, and of antigens of the mouse which include both spermatozoal autoantigens and Ia antigens (Millette and Bellvé, 1977, 1980; O'Rand and Romrell, 1977, 1980; Isojima and Li, 1968; Fellous *et al.*, 1976). Figure 3.6 records the incidence of one spermatozoal autoantigen of the mouse at different stages of spermatogenesis.

The belated appearance of spermatozoal autoantigens, after the formation of an effective isolating barrier, may be necessary to avoid the possibility of a harmful autoimmune response. The late expression of Ia antigens, which are not autoantigenic, does not confer any such advantage. This suggests that the timing of antigenic expression may have some other significance. Besides, some sperm-specific antigens do not conform to this pattern. Figure 3.7 shows the variation in the number of antigenic sites of

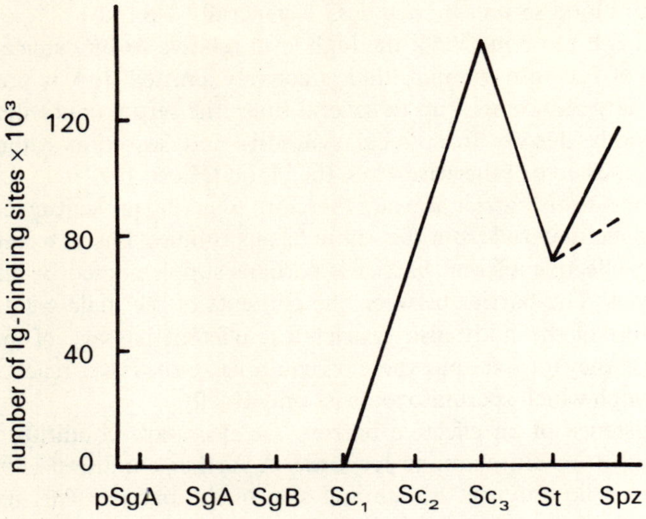

Figure 3.6 Estimated number of antigenic sites per cell at various stages of mouse spermatogenesis. The antiserum was prepared against vas deferens spermatozoa and absorbed before use with mouse thymic lymphocytes and other somatic cells to remove antibody against common antigens. (pSgA = primitive type A spermatogonium; SgA, SgB = spermatogonia type A and type B; Sc_1, Sc_2, Sc_3 = preleptotene, leptotene/zygotene, pachytene spermatocytes; St, Spz = spermatid, spermatozoon. The broken line indicates the residual body.) (Based on data from Millette and Bellvé, 1977.)

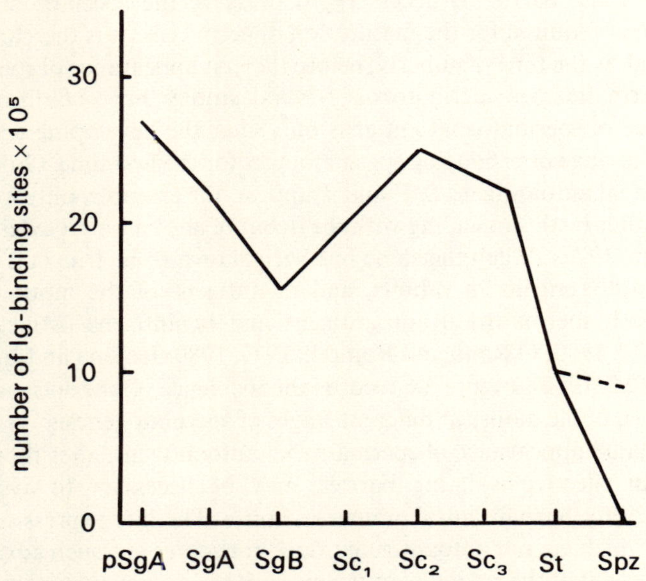

Figure 3.7 As for Figure 3.6; in this case the antiserum used was prepared against type B spermatogonia. (Based on data from Millette and Bellvé, 1980.)

another antigen, shared by mouse germ cells and Sertoli cells, but absent from other tissues of the body. This was detected by antisera prepared against Type B spermatogonia. The final rapid decline in antigenicity of the germ cells with respect to this antigen is due to the partitioning of the antigenic sites into the cytoplasmic membrane of the residual body, which remains in the Sertoli cell when the mature spermatozoon casts off into the lumen of the seminiferous tubule. Possibly this explains the presence of the antigen in Sertoli cells; it may be a truly spermatozoal antigen, detectable in Sertoli cells because of their acquisition of residual body material. The drastic reduction in the number of antigenic sites on each spermatozoon might be relevant to survival in the female genital tract (Millette and Bellvé, 1980).

Circumvention of the blood–testis barrier by injection of semen-specific antigens elsewhere in the body can lead to immunological attack on the normally protected regions of the male genital tract. Of the common laboratory animals, the guinea pig is the most amenable to sensitization to seminal antigens.

As the barrier is less impregnable in the rete testis than in the seminiferous tubules themselves, it is often here that the first signs of immune aggression are apparent after injection of, for example, testis homogenate suspended in an appropriate adjuvant. (An adjuvant is a material which potentiates an immune response without determining its specificity.) Local inflammation develops, with penetration of the epithelium by increasing numbers of leucocytes of various types. Either by reflux of immunoglobulin and invading leucocytes back along the lumen, or by spreading of the disruption of the barrier, the damage extends progressively backwards into the seminiferous tubules. Soon there is considerable oedema of the peritubular spaces and accumulation there of lymphocytes, plasma cells and macrophages. Within the seminiferous tubules conspicuous vacuolation appears first in the spermatids and spermatocytes, later in the spermatogonia. The more advanced germ cell stages exfoliate first and are carried away, down the lumen and out of the testis; eight weeks or so after the initial injection only Sertoli cells and some spermatogonia remain. Antispermatozoal antibody of various types, including cytotoxic and complement-fixing, is detectable in the serum at this stage, accompanying the histological signs of delayed hypersensitivity. Sterility may persist for several months, but may eventually be followed by a spontaneous resumption of spermatogenesis and a return to fertility. This collection of symptoms is referred to as *autoimmune aspermatogenic orchitis* (AIAO); 'orchitis' denotes inflammation of the testis (Andrada *et al.*, 1979; Beer and Billingham, 1976; Muir and Turk, 1979; Tung, 1977; Voisin *et al.*, 1974).

Homogenates of the testes of young males do not elicit AIAO, so the antigens responsible are presumably found only on spermatozoa, or other germ cells such as spermatocytes, which appear only with puberty. Of the specific autoantigens discussed above (pp. 34 and 41), S, P, T and Z are all effective, in purified form, in eliciting AIAO, in doses as low as the equivalent of 5×10^5 spermatozoa. Several different antigens thus produce a common eventual response, although there are differences in emphasis, and of relative time of onset, of the different elements of the AIAO syndrome depending on which antigen is used in sensitizing. With P, the complete reaction is slower to develop than with S, and the delayed hypersensitivity element appears last; with S it develops early in the response. AIAO is therefore complex, with a number of facets which may be differentially evoked by different antigens (Shulman, 1974).

Although humoral antibodies of various types are produced as well as the cell-mediated delayed hypersensitivity response, probably the latter is more important; the severity of AIAO usually correlates better with the intensity of the delayed hypersensitivity than with serum antibody levels, while transfer of regional lymph node cells is necessary for the passive transfer of AIAO to a previously unaffected male. However, in about a quarter of cases autoimmune aspermatogenesis occurs in the absence of any sign of delayed hypersensitivity, with humoral antibody presumably the sole effector (Toullet and Voisin, 1976). The finding that passive transfer can sometimes be achieved with serum alone, together with recent elucidation of the interactions between T and B lymphocytes and macrophages, suggests that AIAO is the result of complex and subtle interplay between cellular and humoral immune mechanisms (Andrada *et al.*, 1979; Brown *et al.*, 1967; Hojo *et al.*, 1980; Tung, 1977, 1980).

If the sensitizing antigen is injected in saline rather than in adjuvant, no delayed hypersensitivity occurs, and the spectrum of antibodies produced does not include cytotoxic antibody. Not surprisingly, no testis lesions develop; in fact the presence of non-cytotoxic antibody protects the testis against any subsequent attempt to induce AIAO by injection of antigenic material in adjuvant. This blocking of a destructive immune response is probably achieved by masking of antigenic sites on germ cells by non-cytotoxic antibody, a mechanism not unlike that involved in the phenomenon of specific immune enhancement (p. 21)—see Shulman (1971).

Similar effects can be induced in other species. Although the nature of the sensitizing autoantigens is less clearly defined, guinea pig sperm antigens do cross-react with those of other species (Rümke, 1974) (p. 36). Autoimmune orchitis similar to that of guinea pigs can be induced by testis homogenates, injected in suitable adjuvant, in mice, rats, rhesus monkeys, bulls, and human volunteers (see Figure 8.1), while in rabbits intense local freezing of one testis may result in an autoimmune attack on the other without the necessity for adjuvant or even artificial injection of antigen (Mancini, 1971; Mancini *et al.*, 1975; Shulman, 1971).

Although ectopic injection of testis material may provoke an autoimmune response, the response to testis autoantigens need not be a damaging one. Indeed, it has been observed (Hurtenbach *et al.*, 1980) that, *in vitro*, autologous mouse testis cells have a propensity to induce suppressor T cells and tolerance, rather than autoimmunity. Although the relevance of this finding to the situation in intact mice is not clear, the finding is of considerable interest.

Injection of male guinea pigs with seminal vesicle fluid, or with a homogenate of seminal vesicle tissue, can result in inflammation of the seminal vesicles with leucocyte infiltration and atrophy of the tubules of the seminal vesicles. This form of autoimmune aggression, analogous to AIAO in the testes, does appear to be rare following the ectopic injection of accessory gland autoantigenic material; a commoner sequel is the appearance of serum autoantibodies with no tissue damage (Orsini and Shulman, 1971; Shulman, 1971).

3.5 Spontaneous autoimmune reactions and infertility

If spermatozoal or seminal plasma autoantigens are removed from the seclusion of the male tract and injected outside the confines of the barrier which normally isolates them, the consequence may be an autoimmune response in which circulating humoral antibody and a cell-mediated response may both participate. Inflammation of the testis, destruction of germinal tissue, and sterility may ensue. Presumably antigen occasionally gains access to the immune system in natural circumstances: is there any evidence for the spontaneous, rather than experimentally-induced, occurrence of autoimmune responses to sperm in male mammals?

The blood serum of male (and female) guinea pigs and rabbits has been reported to contain a natural (that is, not induced) anti-spermatozoal antibody, complement-fixing and capable of causing cytolysis of spermatozoa *in vitro*. In humans, similar natural antibodies against a number of identified spermatozoal antigens have also been identified, the greatest incidence being in prepubertal children, regardless of sex. This is surprising, since spermatozoal antigens would be expressed in boys only at puberty, and in girls not at all. However, since some at least of the antibodies in question react also with common microorganisms such as *Escherichia coli* and *Klebsiella pneumoniae*, the explanation is probably that these—and possibly by extension all 'natural' antibodies reacting against spermatozoa—result from immunization against microorganisms (Tung, 1977).

The accidental cross-reaction between sperm and microorganisms might nevertheless have implications for reproduction. Firstly, immunological tolerance of spermatozoal antigens is precluded if it would also confer tolerance of potential pathogens. Secondly, the prior presence of circulating antispermatozoal antibody might be beneficial, and block the inception of deleterious autoimmune responses against sperm, by destroying stray spermatozoa immediately on their emergence from the blood-testis barrier and before contact is made with the cells of the immune system. This, if it occurred, would be similar to the situation in Rhesus-incompatible pregnancies, where natural antibody of the ABO blood group system may block sensitization to Rhesus antigens on fetal erythrocytes (p. 146).

In humans a non-specific spermagglutinin, which is not an immunoglobulin, has also been described. It is probably a β-lipoprotein of molecular weight somewhat greater than 200 000. This might have a similar role (Boettcher, 1974).

Spontaneous occurrence of antispermatozoal antibody in men (cross-reactions apart) is by no means unusual, and many investigators have endeavoured to correlate its presence with otherwise unexplained infertility. Several types of antibody are involved; those present most frequently, and generally at higher concentrations (or at least more readily detectable) are sperm-agglutinating. They belong to both IgG and IgM classes, the former predominantly causing tail-to-tail agglutination of the sperm, and the latter head-to-head. Tail-to-tail—hence IgG—appears to be of greater importance in men (and head-to-head in women, when spermagglutinins are induced). The difference in action presumably reflects differential sensitivity to antigens regionally distributed on the sperm surface membrane (p. 36). As agglutinating antibody acts on intact sperm, antigens below the surface are unlikely to be implicated (Beer and Neaves, 1978; Hjort, 1977; Rümke, 1974, 1978).

Serum spermagglutinating antibody tends to increase in incidence with age. It is generally absent from boys before puberty, although one nine-year-old has been found to have significant levels. This could be due to precocious development of spermatozoal antigen, or perhaps to some undetected cross-reaction (Fjällbrant, 1975; Tung, 1977). Sperm-immobilizing antibody, of both IgG and IgM classes, cytotoxic and complement-fixing, may also be present in the serum.

Assessments of the incidence of antispermatozoal antibody levels vary according to the techniques used to detect them. Using the fairly stringent criterion of the occurrence of agglutinating activity in blood serum after a 32-fold dilution (that is, five serial dilutions of 1:2), about 3% of infertile men show the presence of serum spermagglutinins, compared with none of a control group of men of proven fertility. This incidence might not seem strikingly high, but is in part a function of the chosen method and criterion of assessment; other methods give somewhat different results. Apart from this consideration, there are obviously many causes of infertility with no immunological significance. Table 3.1 shows the relative incidence of sperm-agglutinating and sperm-immobilizing activity in the serum of fertile and infertile men, as recorded in one particular survey.

Although there is some correlation between the presence of anti-

Table 3.1 Incidence of sperm-agglutinating antibody in the serum of fertile and infertile men (data from Fjällbrant, 1968*a*)

	Antispermatozoal agglutinin concentration		
	Negative	*Low* (1:2^5)	*High* (1:2^5)
Infertile	373 (93.3%)	10 (2.5%)	17 (4.3%)
Fertile	487 (97.4%)	9 (1.8%)	4 (0.8%)

spermatozoal antibody and infertility, and between the serum level of such antibody and the probability of conception (Figure 3.8), it is clear that the presence of such antibody does not necessarily entail infertility. Nor, even in infertile men, is antisperm antibody when present necessarily the cause of infertility. For antibody to form, presumably the integrity of the blood–testis barrier must be compromised. Whatever causes this, or prevents sufficiently rapid elimination of the antigenic material to block the inception of an autoimmune response, may itself be the direct reason for the infertility, with antibody formation a purely incidental secondary result.

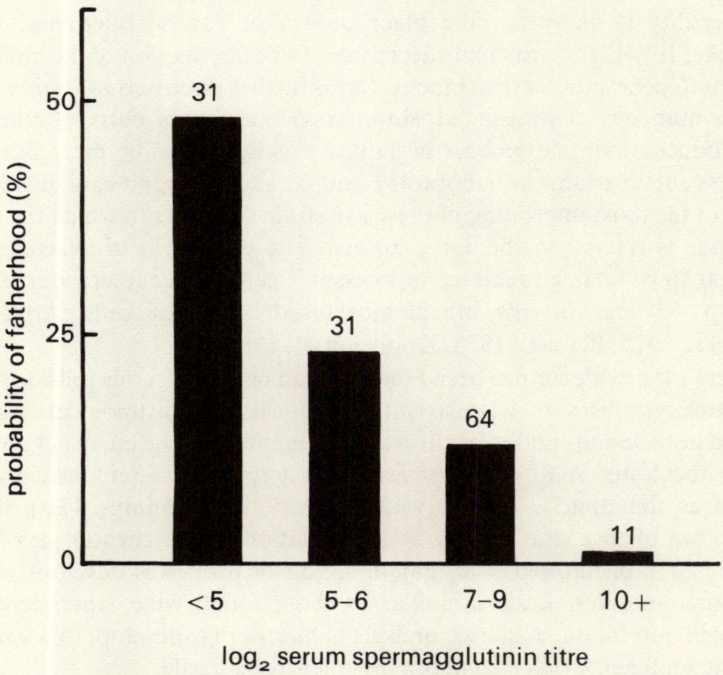

Figure 3.8 Proportion of men with various serum spermagglutinin concentrations who subsequently became fathers. The figure at the head of each column represents the number of men in that category. (Based on data from Rümke, 1978.)

Physical obstruction of ducts of the male genital tract appears to result in the escape of sperm antigens and sensitization against them: a high proportion (50–60%) of men develop antispermatozoal antibodies after vasectomy. In the circumstances, infertility can hardly be attributed to immunological factors (although the presence of circulating antisperm antibody becomes a problem following attempts to reverse sterilization by reanastomosis of the vas). Less overt causes of infertility than vasectomy might also lead to fortuitous antibody formation (Boettcher, 1977; Rümke, 1980).

As the seminiferous tubules are normally impermeable to immunoglobulin molecules, antispermatozoal antibody in the blood has no effect on spermatogenesis. Antibody is, however, present in significant amounts in the seminal plasma of the ejaculate, having entered mainly in the secretions of the prostate. In infertile men, the concentration of antispermatozoal antibody in semen may even approach that in blood serum. Since antibodies as a whole are present at only a fraction of their serum levels (p. 41), this suggests that there is preferential passage of antispermatozoal antibody compared with antibody of other specificity.

As antibody and spermatozoa are brought into contact only when prostate secretions are added during ejaculation, any effect of the antibody on fertility is likely to take place only after coitus (Boettcher, 1978; Rümke, 1974). The principal effect seems to be a reduction of the ability of sperm to penetrate cervical mucus; this is further discussed in section 4.4.

Spontaneous cell-mediated autoimmune responses corresponding to the induced immune orchitis of guinea pigs appear to be rare, although some genetic strains of laboratory animal show a significant incidence. One of the most interesting of these is a strain of mouse in which the testis damage is related to the T/t genotype. The germ cells of affected mice appear to be unable to induce suppressor T cells (hence tolerance) *in vitro* (see p. 40) and, *in vivo*, are phagocytosed by Sertoli cells (Artzt and Bennett, 1975; Bennett, 1975; Dooher *et al.*, 1981).

Very little evidence has been found in man of spontaneous cell-mediated immune responses to testicular antigens, nor is there histological evidence of the testis lesions and infertility which might follow the escape of antigen from the testis. A form of orchitis with a rather different origin does, however, sometimes accompany adult infection with mumps. The primary symptom of this viral disease is inflammation of the parotid glands. In about 20% of infected men, inflammation of the testes develops as the parotitis subsides. Lesions similar to those found with experimentally-induced autoimmune allergic orchitis of guinea pigs develop, to a varying extent, and men so affected may subsequently be sterile.

Mumps orchitis is probably the result of sensitization to an antigen common to both parotid tissue and spermatozoa; strictly, 'organ-specific' to neither. It does not occur in boys infected with mumps before puberty.

Although the antigen is present in the parotid glands, it is presumably one of the antigens restricted to spermatozoa or to post-meiotic germ cells produced only after puberty, so that sensitization to parotid antigen before puberty has no effect on the testis (Beer and Billingham, 1976).

CHAPTER FOUR

REPRODUCTION AND IMMUNITY IN THE FEMALE

4.1 Reproduction in female mammals

Before considering the immune capability of the female reproductive system, it is as well briefly to summarize the relevant anatomy and physiology of reproduction in female mammals. Figure 4.1 illustrates the structure of the female genital tract in two representative species of mammal.

The paired ovaries have a dual role, as endocrine organs as well as the source of gametes, and a correspondingly composite structure. Extremely

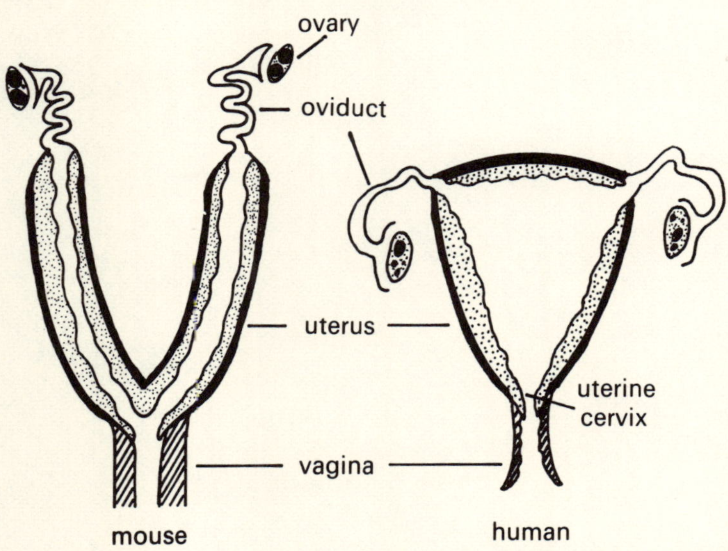

Figure 4.1 Diagram of the female reproductive tract of a typical rodent such as the rat or mouse (left) and of a woman (right).

large numbers of germ cells (potential ova) are present, in varying states of maturity. The earlier stages, oogonia, are diploid and represent a proliferative, mitotic, phase of development. By about the time of birth in most mammalian species, the majority of oogonia have entered the initial stages of meiosis as oocytes, and it is as haploid oocytes that they will eventually be ovulated.

The somatic component of the ovary comprises a heterogeneous collection of cells of different types. During its development, each oocyte is invested with a layer of follicle cells which, as maturation proceeds, multiply and differentiate into several histologically distinguishable layers. The cells immediately surrounding the oocyte are known as *granulosa* cells, and the layer distal to this as the *theca* (further subdivided into *theca interna* and *externa*). Between granulosa cells and oocyte lies a noncellular layer, largely composed of protein and mucopolysaccharide, the *zona pellucida.*

Before puberty, many follicles start with the processes of growth and maturation; only after puberty does a minute fraction of these follicles contrive to complete the process and attain ovulation. As final maturation proceeds, the follicle becomes distended with fluid which accumulates in the granulosa layer. Eventually, rupture takes place, and the oocyte is released; at this stage the zona pellucida is intact, and surrounded by a large number of more or less loosely attached follicle cells, known as the *cumulus oöphorus.*

The newly ovulated oocyte is wafted down the oviduct, or Fallopian tube, by a combination of ciliary and peristaltic action. While it is still in the upper reaches of the oviduct, a spermatozoon may succeed in penetrating cumulus and zona, and achieving fertilization. The zygote

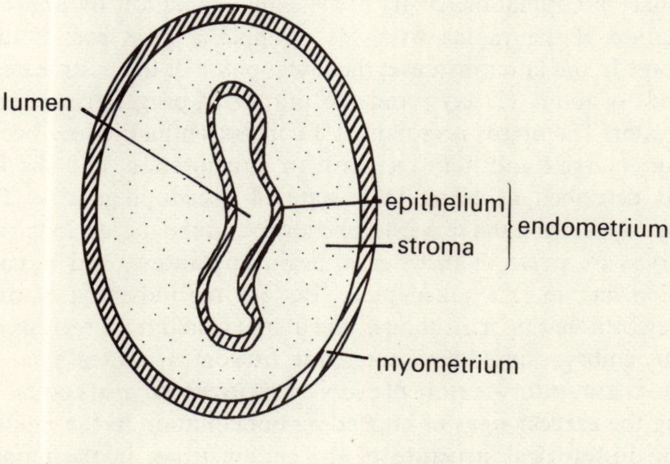

Figure 4.2 Transverse section of the uterus of a mammal.

continues on its way towards the uterus, at some stage shedding the zona pellucida and the residual cells of the cumulus.

On arrival in the uterus, some days after ovulation, implantation may proceed, but only if the endometrium of the uterus (Figure 4.2) has been suitably prepared by ovarian hormones. The hormones in question are steroids: before ovulation, mostly oestrogen, contributed by thecal cells of the ovary. The endocrine function of the ovary (and its gametogenic function) are under the control of the anterior pituitary gland which produces, as gonadotrophic hormones, the glycoproteins *follicle stimulating hormone* (FSH), *luteinizing hormone* (LH) and *prolactin.* These in turn are under the control of the hypothalamus, the release of the first two being jointly determined by an oligopeptide releasing factor known as *LH-releasing hormone* (LHRH), among other acronyms. Prevailing levels of these hormones are regulated by a complex and interacting series of feedback relationships which need not be discussed here (Austin and Short, 1972*c*, 1979; Hogarth, 1978; Johnson and Everitt, 1980).

Once ovulation has taken place, the granulosa cells remaining within the empty follicle hypertrophy and form a structure known as a *corpus luteum* (CL); somewhat misleadingly, since in most species the name ('corpus luteum' means 'yellow body') belies its actual appearance. The corpus luteum is an endocrine organ, whose major product is progesterone. This is essential for the preparation of the uterus to receive an implanting embryo, and for its maintenance as a favourable site for early development.

The fluctuations in hormone levels which occur in the course of normal reproductive cycles in mice and humans are shown in Figure 4.3, where the common features and differences are apparent. In an infertile mouse oestrous cycle, the period of activity of the corpora lutea is brief and their output of progesterone unimpressive. Full and prolonged luteal activity is conditional on copulation, or its convincing simulation by appropriate manipulation of the vagina with, for example, a glass rod of suitable dimensions. In the latter instance, the luteal phase of the oestrous cycle is prolonged to about 12 days, and the output of progesterone becomes much greater. The uterus responds as if conception had indeed occurred, becoming enlarged and fully receptive to implantation, and the female mouse is described as being in a state of pseudopregnancy. This is necessary to maintain the progesterone level, and the uterus, long enough for embryos to arrive in the uterus, begin implanting, and signal that fertilization has, in fact, taken place. But for the induction of pseudopregnancy by a mating male mouse, the uterus would have regressed long before any embryos could have entered it. In women, a luteal phase of 12 days or so is a statutory feature of every cycle, irrespective of coitus.

Among the earliest signs of impending implantation in the mouse is a change in histological structure of the endometrium in the immediate vicinity of each implanting embryo. The fibroblast-like cells of the stroma

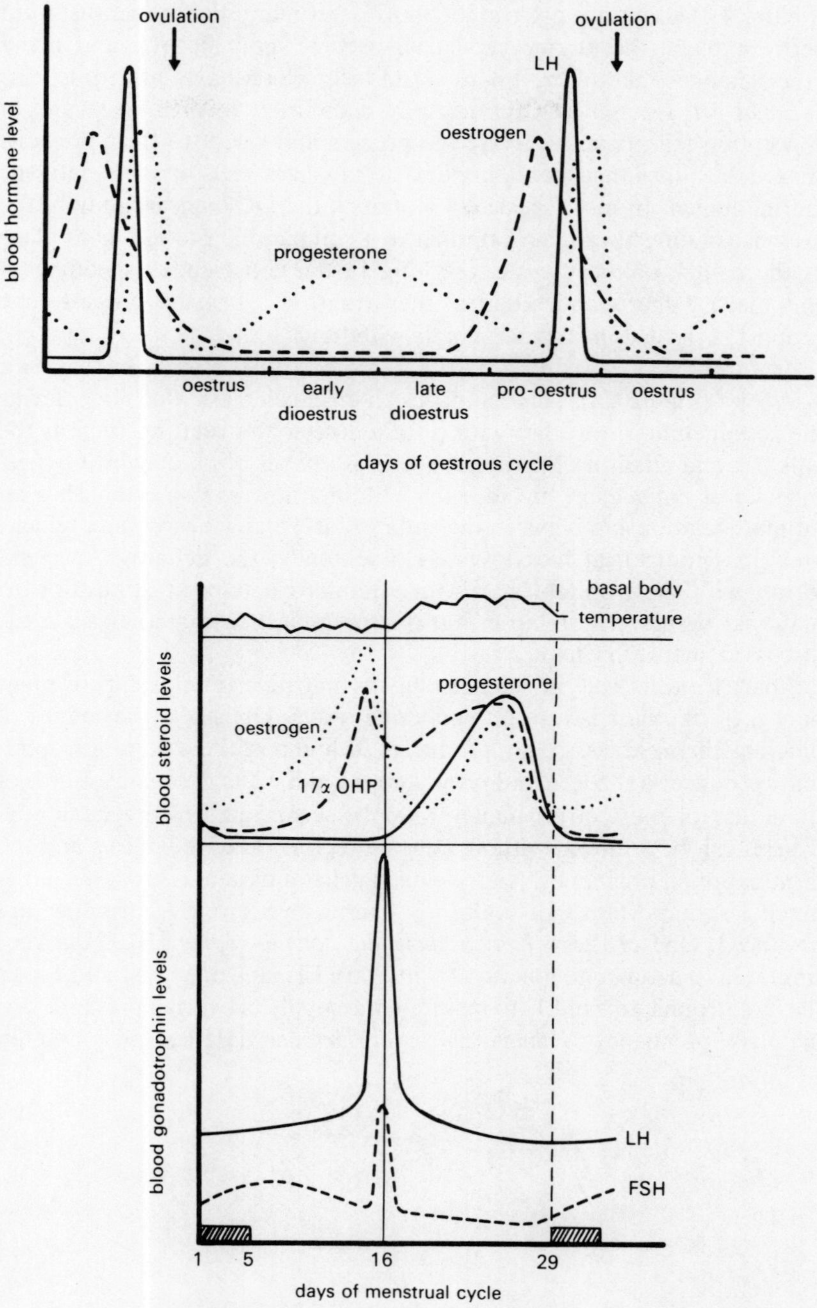

Figure 4.3 Hormonal fluctuations in the course of the oestrous cycle of a mouse (*a*) and in the human menstrual cycle (*b*) (redrawn from Austin and Short, 1972*c*).

(Figure 4.2) adopt a more rounded form and many tight junctions form between them, the uterine capillaries become more porous, and many other changes take place. Endometrial cells which have undergone this *decidual cell reaction* (DCR) are fully committed to pregnancy and, if conception fails, cannot reverse the process and revert to their previous state. Instead, as their name implies, decidualized cells are shed into the uterine lumen. In most species of mammal, the DCR occurs only in the presence of an embryo, but in primates a comparable process takes place in the course of every cycle, resulting in the conspicuous shedding of endometrial debris and blood at menstruation, when the progesterone support of the endometrium is finally withdrawn.

At the time of implantation, the embryo is in the form known as a *blastocyst* (Figure 4.4). This comprises an inner cell mass which later forms the actual embryo together with certain accessory structures such as the yolk sac and allantois. Surrounding this is a layer of tissue which hypertrophies as *trophoblast*, invades the endometrium and so establishes an intimate relationship between the embryo for which it serves as a vehicle, and the endometrial blood vessels. Eventually, the definitive placenta forms, with the trophoblast still maintaining the closest apposition to maternal tissues. The relationship of trophoblast to maternal tissues is discussed further in Chapter 5.

Apart from its role in establishing the nutritive relationship of pregnancy, trophoblast is also an endocrine tissue. Unusually, hormones of different chemical classes are produced. In humans, the placental trophoblast produces oestrogen and progesterone, and is the dominant source of the latter for most of pregnancy. For the second half of pregnancy the ovaries can be removed without causing regression of the uterus and the termination of pregnancy. (In rodents in general placental progesterone is much less important.) In addition, several glycoprotein hormones are produced. One of these, *human chorionic gonadotrophin* (HCG) is very similar in structure and function to pituitary LH, and appears to take over the luteotrophic role of LH to maintain the supply of ovarian progesterone in early pregnancy. *Human placental lactogen* (HPL) (alias human

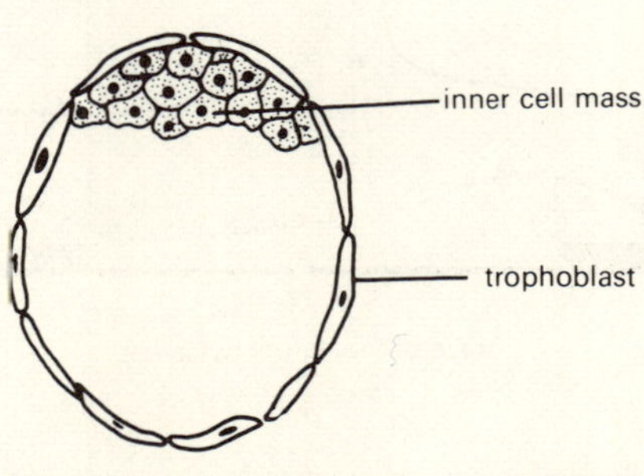

Figure 4.4 Diagram of a mammalian blastocyst.

chorionic somatomammotrophin, HCS) is also produced, in rising amounts, as pregnancy progresses. These hormones may, in addition, have immunological effects. This question, and the crucial role of the trophoblast and placenta in the survival of the fetus in an immunologically hazardous environment, are discussed in Chapters 5 and 6.

From the point of view of the spermatozoa, the female genital tract appears 'more like an obstacle race track than a device to promote egg–sperm union' (Cohen, 1969). In humans, semen is ejaculated into the vagina. Before entering the uterus, sperm must pass through the narrow cervix of the uterus, which is filled with cervical mucus. For most of the menstrual cycle, this mucus renders the cervix impassable, to sperm or to any other extraneous materials; only around the time of ovulation does the consistency of the mucus alter in such a way as to allow live sperm to enter. Dead or abnormal sperm, and most foreign particulate matter, are still selectively excluded; even the bulk of the seminal plasma is left behind in the vagina by the departing sperm.

In mice, most of the ejaculate passes very rapidly into the uterus. In many species where semen is deposited in the uterus, the major barrier to sperm is a valve-like uterotubal junction which controls the passage of sperm from uterus to oviduct.

The effect of these restrictions on the upward passage of sperm is that only a tiny fraction of the total sperm in an ejaculate get anywhere near an egg. In female rabbits, for instance, there are approximately 2×10^6 sperm in the region of the uterine cervix 12 hours after mating, and 4.4×10^6 in the uterus, but only 1.1×10^4 in the lowest stretch of the oviduct, and a meagre 60 or so surrounding the actual oocyte in the upper reaches of the oviduct. All other sperm die, or are killed, shortly after mating, and are destroyed by leucocytes of various types. The sperm which have the opportunity of fertilizing eggs may not be a random sample of the sperm in the ejaculate, but may have been selected on the basis of their phenotypes. As the sperm within an ejaculate can be antigenically heterogeneous, selection on an immunological basis is at least a possibility.

4.2 Immune response in the female genital tract

Anatomically, the immune system relating to the female reproductive tract is similar to that elsewhere in the body. The requisite afferent vessels and regional lymph nodes draining the uterus are present (Figure 4.5), the efferent arm of the immune system being represented by the uterine and ovarian arteries and capillaries. Lymphoid cells, including plasma cells, are present within the mucosal layers of the genital tract, as within the corresponding mucosae of gut and bronchus, although compared with these organs the distribution is sparse.

Humoral antibody is present at all levels of the female reproductive

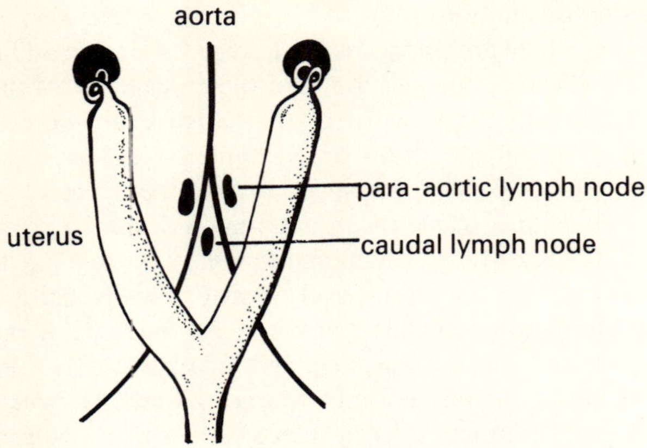

Figure 4.5 Position of the principal regional lymph nodes relevant to the uterus.

system. Follicular fluid apparently represents a filtrate of blood serum, selective only with respect to simple properties of serum components, such as molecular size, smaller molecules crossing the basement membrane of the follicle more readily than larger ones. Within human follicular fluid, IgG, IgA and IgM are all present, in concentrations, respectively, of 45%, 36% and 12% of those obtaining in serum. The descending order of concentration, relative to blood serum, reflects the ascending order of molecular weight (Table 2.1, p. 12). Other species show roughly similar proportions of the major immunoglobulin classes (Beller and Schumacher, 1980; Cinader and de Weck, 1976; Symons and Herbert, 1971).

Ovulation discharges this fluid, along with oocytes, into the oviduct. Although the absolute amounts of antibody present in follicular fluid are not large in relation to the total volume of the lumen of the reproductive tract, the concentration in the immediate vicinity of the oocyte may, for a short time at least, be high, particularly among the cumulus cells through which a hopeful spermatozoon must pass. Human blood group antibodies are among those whose presence has been detected in follicular fluid, and sperm carrying the corresponding antigens (section 3.2) might be at a disadvantage.

The relatively direct access of large molecules from blood is not a feature typical of the tract as a whole, and in general large molecules such as immunoglobulins are found in only low concentrations elsewhere. In the rabbit, immunofluorescence studies of the distribution of IgG suggest that the epithelial layer is the main barrier to free access of immunoglobulin from blood serum to lumen at all levels of the genital tract, but particularly in the uterus. Much the same is probably true of other species. Nevertheless, immunoglobulin of various classes is detectable within the luminal secretions, although often in amounts too small to measure accurately. In the oviduct of the cow, quite large amounts of IgG1 and IgG2, and small amounts of IgA, have been found. IgG, IgA and IgM have been shown within the villi of the human oviduct, and the 'secretory piece', associated with IgA (p. 11) has also been detected in the epithelium of the oviduct.

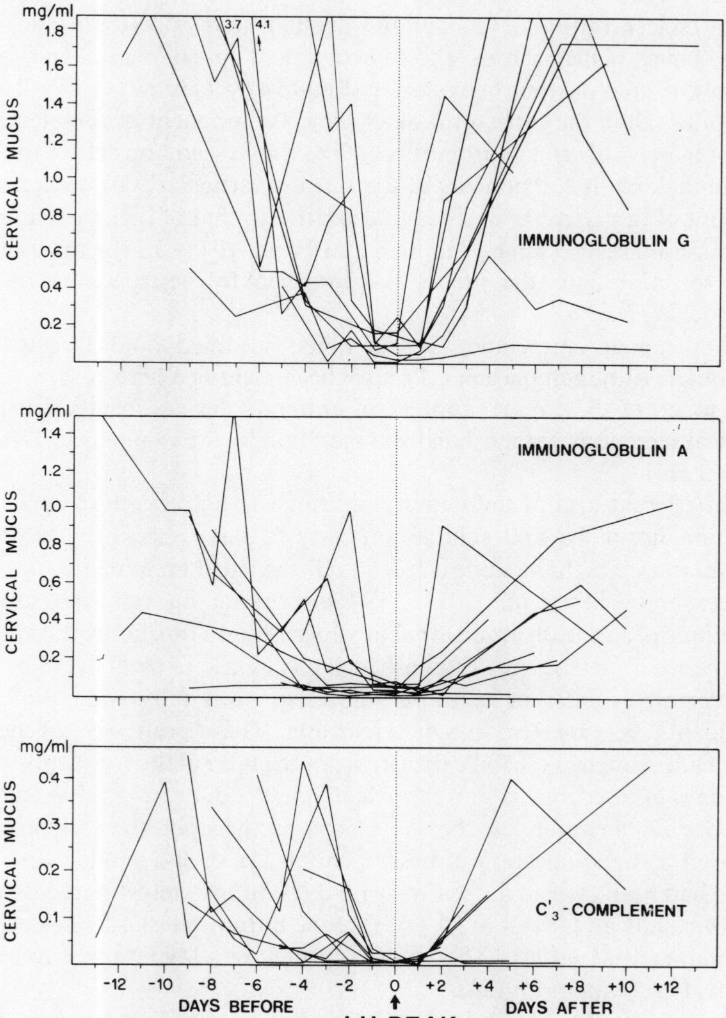

Figure 4.6 Levels of IgG, IgA and of the C3 component of complement measured in the cervical mucus of women around the time of the ovulatory LH peak (from contribution by Schumacher, G. F. B. to discussion of Vaerman and Férin, 1974).

The same three immunoglobulin classes are also represented in uterine fluids.

Cervical mucus is more easily sampled free from contaminants than the secretions of higher levels of the female tract, so correspondingly more is known of its composition. In human cervical mucus, the major immunoglobulin present is IgG (up to 6 mg ml^{-1}), with IgA concentrations somewhat lower (up to 1.4 mg ml^{-1}), and IgM in only trace amounts (Cinader and de Weck, 1976; Parish *et al.*, 1967).

The concentrations of IgG and IgA fluctuate in relation to the menstrual cycle, being higher during the follicular and luteal phases and falling sharply around the time of ovulation (Figure 4.6). This mid-cycle fall—and a parallel fall in the concentration of the C3 component of complement—might merely reflect a dilution effect of the rise in secretion rate of the non-immunoglobulin components of the mucus, particularly of water; as the amount of IgA rises about three-fold relative to that of IgG, the pattern of secretion must also alter (Vaerman and Férin, 1974). In the rat, IgA and IgG secretion into the uterus is controlled by oestrogen (Wira and Sandoe, 1977).

The vaginal epithelium is squamous, stratified, and largely non-glandular. Although plasma cells have been identified here, it seems likely that much of the vaginal content of antibody derives from uterine and cervical secretions, rather than from secretion by the vagina itself (Rebello *et al.*, 1975).

The efferent arm of the humoral immune system therefore appears to function normally, and soluble antibody to have access to the female genital tract. Much the same is true of cell-mediated immunity.

Beer and Billingham (1974*a*, 1976), working on rats, developed a technique of surgically inserting skin allografts into the uterus through the uterine wall. Firstly, they established the response to grafts of skin from syngeneic donors; that is, those whose skin was fully compatible and would not be expected to suffer rejection. These grafts were generally physically sloughed off and extruded from the uteri of the recipients within two days of insertion; in one experiment, only 2 out of 24 grafts remained in place. If, however, on the same days as the graft the recipient also received a single injection of oestrogen (50 μg), or if a similar oestrogen surge had been evoked naturally 4 or 5 days prior to insertion of the graft, then virtually all (48 out of 50, 96 %) of the intrauterine grafts healed in to the endometrial surface. The uterus is therefore a favourable site for the survival of compatible grafts.

The fate of allografts in females which had previously been immunized by conventional methods contrasts strongly with this. Whether pregnant or pseudopregnant (from mating, or artificial cervical stimulation 4 or 5 days previously) no intrauterine graft survived longer than 5 days. A pre-immunized female is therefore fully capable of rejecting intrauterine allografts; presumably sensitized cytotoxic T cells have no difficulty in entering the uterus, and the efferent arm of the cell-mediated immune system is fully functional there.

An afferent block to allograft rejection might, however, allow survival of an intrauterine allograft; that is, an allograft would survive if sensitization could not be induced from within the uterus, and if immunity had not previously been induced by extra-uterine means. It is possible, in other words, that the uterus constitutes an 'immunologically privileged site' where an allograft will fail to evoke an immune response despite being at the mercy of one induced elsewhere. Can allograft rejection be initiated, as well as expressed, in the uterus?

Immunologically privileged status might result from a lack of afferent drainage of the uterus. In the rabbit uterus, whereas the myometrium is liberally supplied with lymphatic vessels, the only endometrial lymphatics are restricted to layers very close to the boundary with the myometrium. Towards the lumen, the endometrium does indeed appear to be deficient in lymphatic drainage. Skin allografts to the endometrium appear to survive in a healthy state slightly longer than those elsewhere (McLean and Scothorne, 1972).

Even this modest privilege is denied to the rat uterus. Allografts to the endometrium of females which had not been presensitized (in contrast to the previous experiment) are rapidly destroyed, with rejection times little different from those of genetically equivalent grafts elsewhere. Syngeneic grafts, as already mentioned, survive indefinitely provided certain hormonal conditions are met.

A sensitive indicator of the occurrence of an immune response within the uterus is provided by the increase in size of the relevant regional lymph nodes, in the case of the rat the para-aortic lymph nodes (Figure 4.5). Figure 4.7 shows the dramatic increase in weight of the para-aortic lymph nodes which follows intrauterine allografts of skin. Similar isografts have negligible effects.

Considerable changes take place in the endometrium in the course of the oestrous or menstrual cycle, and of pregnancy. Although a uterus primed with oestrogen might accept isografts and reject allografts, it is still possible that the histological changes which occur with decidualization of the endometrium and the establishment of pregnancy might impede afferent lymphatic drainage and so permit the survival of allografts.

This was tested by transferring allografts of tissue into the uteri of female rats which had been mated syngeneically four days before. The tissue grafts would thus be healing in to the endometrial surface at about the same time as the blastocysts resulting from the mating were evoking a decidual cell reaction and implanting (p. 54). In these circumstances, allograft survival *was* extended, and most allografts appeared healthy even after 17 days. Much the same was true of grafts into the uteri of pseudo-

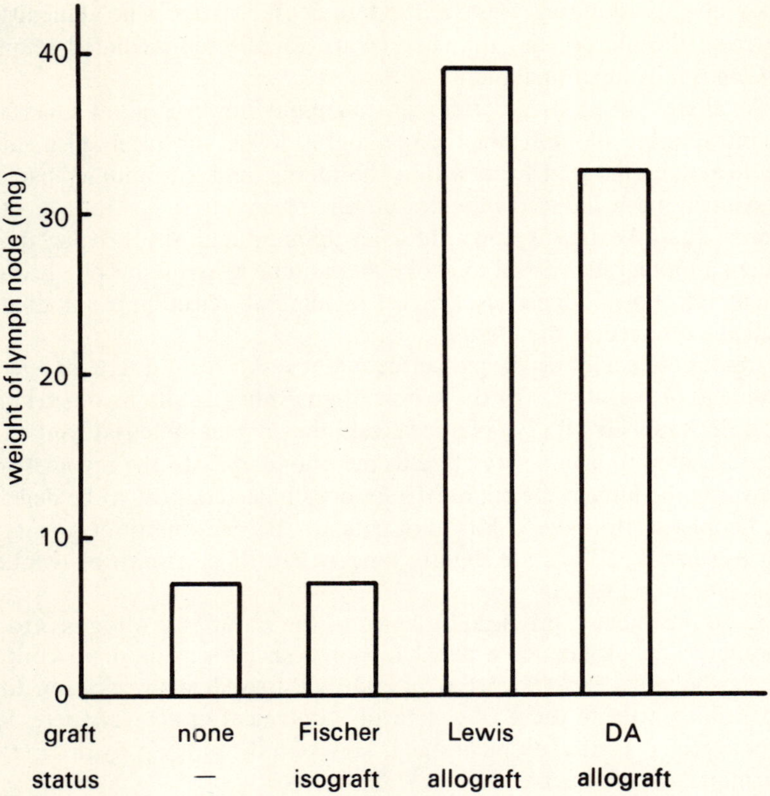

Figure 4.7 The effect of intrauterine allografts of skin on the weight of the para-aortic lymph nodes draining the uterus. In each case the recipient animal was a rat of the Fischer strain. Graft donors were of either Fischer, Lewis or DA strains; Fischer and Lewis strains were incompatible only at minor histocompatibility loci (based on data from Beer and Billingham, 1974*a*).

pregnant females. The precise reason for this is not clear; apart from simple failure to detect alloantigen, one possibility is that the DCR, by reducing afferent lymphatic drainage, diverts antigen from the lymphatic system more directly into the blood. In other contexts this is known to bias the immune system in favour of producing 'blocking antibody' which impedes a potentially destructive cellular response without itself being cytotoxic (p. 22). Whatever the precise explanation, it does appear that in some circumstances the decidualized uterus affords some protection to allografts.

In addition to its participation in general forms of immune response, the female genital tract may also have a form of locally-mediated immune response. Local immune responses are a feature of mucosal surfaces, notably those of the gut and bronchi, whose secretions contain significant amounts of antibody. Mostly, secretory antibody is IgA, but IgE and IgM have been identified as well (Bienenstock and Befus, 1980; Roitt, 1980).

Secretory antibody may be transferred from the blood serum, or synthesized locally. In the latter case, the plasma cells in question might have infiltrated the mucosal epithelium from elsewhere in the body, or might have originated in response to local activation of lymphocytes by local antigenic stimulation. In most cases the available evidence does not conclusively indicate which of these mechanisms applies.

Where lymphoid cells (particularly plasma cells) are sparse or absent, synthesis within the mucosa is unlikely. Some workers report a sparse population of lymphoid cells in the endometrium of the human uterus, for instance, others a total absence (Cinader and de Weck, 1976). IgA and the 'secretory piece' associated with it are both present within epithelial cells. It seems likely, therefore, that the endometrium lacks significant ability to synthesize antibody, but produces its own secretory piece material. As this has an affinity for the J-chain of IgA, it could then act by accumulating IgA (in the dimer form) from the blood, and transferring it to the luminal surface for release. This accumulation and transfer could explain why the concentration of IgA in secretions of the uterus is often greater than that in blood serum (Kelly and Fox, 1979).

The epithelium of the uterine cervix also contains secretory piece material, but also has fairly abundant IgA- and IgG-secreting plasma cells. These increase in number during local infections of the genital tract, so the cervix probably does have a true local immune system. In contrast again, the human vagina lacks plasma cells and is lined with a squamous epithelium, unsuitable for secretion. Other species differ from humans: for example, organ culture of bovine endometrial and vaginal tissue shows that both are capable of IgA synthesis.

The importance of locally-produced IgA is hard to assess; although present in small amounts, it might conceivably modulate in some way the access of antigen to IgG-producing, or other, lymphoid cells of the general immune system.

One particularly interesting manifestation of local cell-mediated immunity in rats has been studied by Beer and Billingham. These workers used female rats of the Fischer strain as recipients; in each case they inoculated one horn of the uterus with a suspension of about 4×10^6 live lymphoid cells obtained from donors of the allogeneic DA strain. The other horn of the uterus received a similar inoculum of syngeneic cells. Loss of the inoculated cells by way of the vagina, or their transfer between uterine horns, were prevented by temporarily ligating the base of each horn.

Three weeks later, each horn was presented with a challenge consisting of a similar cell suspension; either of the same strain as previously, or of the other strain. A uterine horn which had initially received allogeneic cells, challenged with similar allogeneic cells, underwent striking enlargement and, within a day or two, became inflamed and distended with fluid. Usually this 'recall-flare' reaction subsided within a further day or so. Broadly speaking, this resembles a form of delayed hypersensitivity (p. 15), known to be a manifestation of cellular, rather than of humoral, immunity.

Only when an initial allogeneic stimulus was followed by a similar allogeneic challenge to the same horn did a 'recall-flare' reaction occur; the reaction induced in one horn remained local and did not spread to the other horn of the uterus. This emphasizes the completely local nature of the response (Beer and Billingham, 1976).

The uterus therefore does have its immunological idiosyncrasies, but despite these intrauterine allografts are decisively rejected in most circumstances. It does not seem likely that spermatozoa or fetal tissues could survive simply because the uterus protects its contents in some way from immune attack; immune responses can be initiated as well as effected within the uterus. Some other explanation of sperm and fetal survival must be sought.

4.3 Spermatozoa and semen within the female genital tract

Semen is antigenic, as demonstrated by a variety of conventional immunological techniques (section 3.2). It contains autoantigens, peculiar to spermatozoa (and their immediate forerunners) or to seminal plasma, and alloantigens including those of the major histocompatibility systems; as well as xenoantigens, antigenic only in other species. The latter are irrelevant to the question of an immune response to normal mating, but both alloantigens and autoantigens might evoke an immune response in the female; as described in the last section, the female genital tract is in general fully capable of both cellular and humoral immune responses of various types. The response to seminal antigens—as opposed to responses to any other foreign antigens—must be blocked, or at least muted so that spermatozoa can survive long enough for fertilization to take place; and

Table 4.1 Response of draining lymph nodes of the uterus to the intrauterine inoculation of syngeneic and allogeneic spermatozoa; effects on survival time of subsequent skin allografts. (Allogeneic spermatozoa and skin from rats of Lewis strain, recipients of Fischer strain.) (Data from Beer and Billingham, 1974*a*.)

	Lymph node weight (*mg* ± *s.d.*)		
Sperm dose	*Syngeneic sperm*	*Allogeneic sperm*	*Mean survival time of graft* (days ± s.d.)
0	7.04 ± 3.60		11.0 ± 1.00
1×10^6	6.3 ± 2.5	20.3 ± 5.7	9.43 ± 0.79
10×10^6	10.8 ± 3.3	23.0 ± 6.0	7.75 ± 0.71
50×10^6	9.3 ± 4.3	33.2 ± 18.5	6.6 ± 0.69

however the female genital tract connives at sperm survival, it must continue to do so not just for a single brief irruption of spermatozoa, but generally for many, over a period possibly of many years. The relative rarity of female infertility of immunological origin testifies to the success with which sperm apparently avoid the natural consequences of their own antigenicity.

There is much evidence to suggest that survival of spermatozoa is not due to a simple failure of the female genital tract to recognize their presence. When washed epididymal spermatozoa are collected from male rats of one genetic strain and deposited into the uterine lumen of females of a different strain, the 'sentinel' para-aortic lymph nodes (Figure 4.5, p. 56) show considerable enlargement, similar to that which results from placing in the uterus an allograft of skin or suspension of allogeneic lymphoid cells. This indicates that the intrauterine inoculation of spermatozoa has sensitized the females to spermatozoal antigens (Table 4.1; compare Figure 4.7). Similar conclusions follow from comparable experiments on mice and hamsters (Beer and Billingham, 1974*a*, 1976).

Controls in which the sperm donors are males of the same strain as the recipient females do not show lymph node hypertrophy, demonstrating that the sensitizing antigens must be strain-specific alloantigens, not

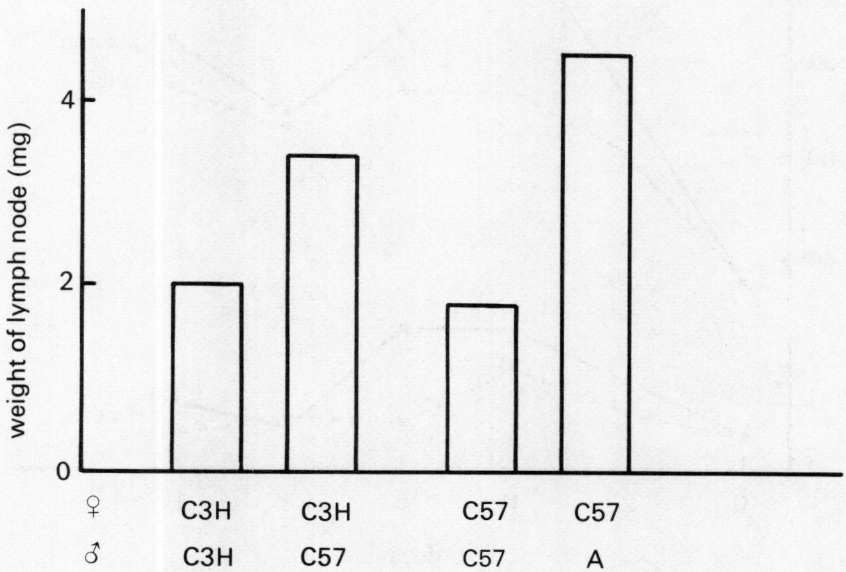

Figure 4.8 The effect of the first mating on the weight of the para-aortic lymph nodes of female mice. One to two hours after mating, the uteri were ligated to retain an estimated $10–50 \times 10^6$ spermatozoa. In both cases the intrastrain matings were MHC-incompatible (based on data from Beer and Billingham, 1974*a*).

antigens peculiar to spermatozoa. When challenge grafts of skin from animals of the same strain as the sperm donors are transplanted some three weeks later, they are rejected more rapidly than similar allografts to females which had not previously received sperm (Table 4.1). This confirms that the sensitizing antigens are in fact alloantigens, common to spermatozoa and skin cells.

The spermatozoa used in these experiments were obtained directly from the epididymis of the donor rats, separated from the seminal plasma, and washed (thus clearing them of sperm-coating antigenic material?) before being injected into the recipient uteri. Spermatozoa do not normally enter the uterus through the agency of a hypodermic needle: are there any immunological consequences of normal insemination?

The 'recall-flare' reaction of the rat, a criterion of sensitization having

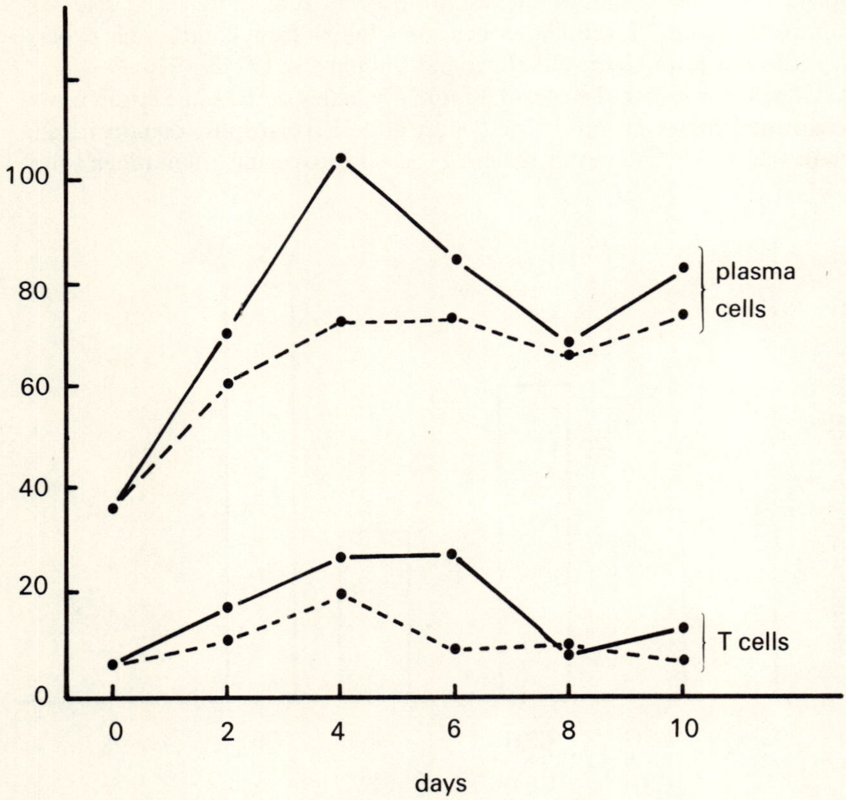

Figure 4.9 Variations in the relative numbers of plasma cells and T lymphocytes in the lymph nodes draining the uteri of female rats in the period following a first interstrain (broken lines) or intrastrain (solid lines) mating. The vertical scale refers to the number of cells counted in 5 randomly-chosen areas of 0.02 mm^2 in 7 μm sections of the lymph node cortex (redrawn and adapted from McLean *et al.*, 1980).

occurred (p. 62), can also be elicited by natural mating, although mating is not in itself a sufficient stimulus to induce the necessary local sensitization in the first place.

Even a single copulation in mice appears to be sufficient to cause hypertrophy of the lymph nodes draining the uterus (Figure 4.8), while in rats it has been shown that proliferation of both T and B cells within the relevant lymph nodes is detectable within two days of the first mating of previously virgin mice (Figure 4.9)—see McLean *et al.* (1980). In both cases the response is at least partly due to alloantigens, since the effects are greater with allogeneic matings.

In contrast to artificial insemination with washed epididymal spermatozoa, natural mating seldom seems to affect the survival time of challenge allografts of skin. There have, however, been reports of mating-induced tolerance to isografts. In certain strains of mice, females are particularly

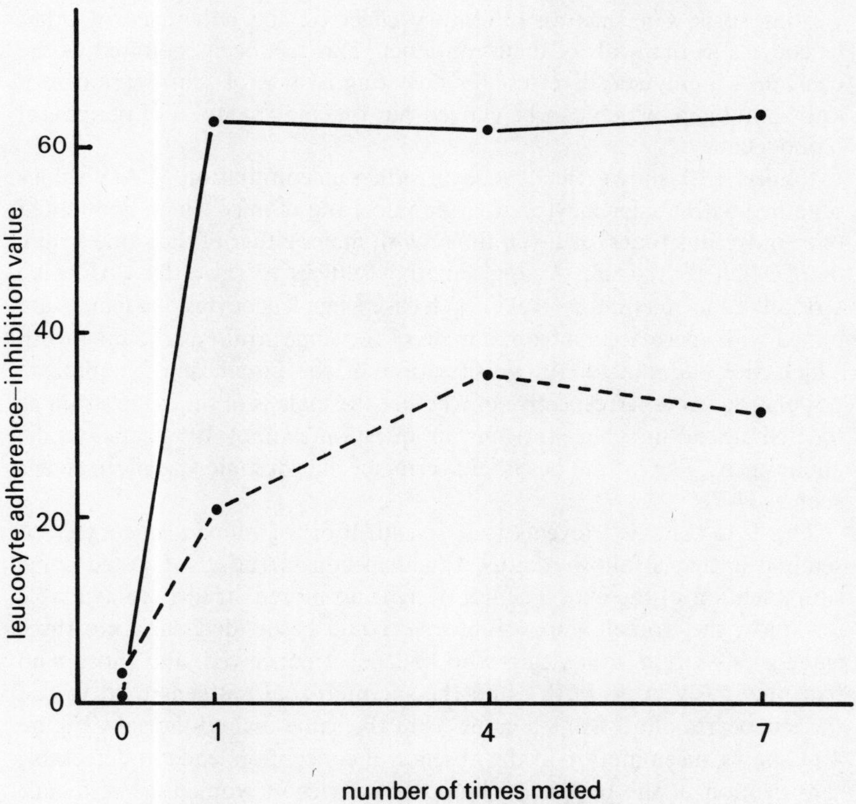

Figure 4.10 Leucocyte adherence-inhibition (LAI) values for virgin female mice and for females mated once, four times, and seven times. Pregnancy was prevented hormonally. Different concentrations of sperm were used for the *in vitro* test of T cell response: 5×10^4 sperm ml^{-1} (broken line) or 20×10^4 sperm ml^{-1} (based on data from Dorsman *et al.*, 1978).

liable to reject grafts of skin from males of the same strain as themselves (Prehn, 1960). The antigen to which the females respond is, of course, the male-specific antigen H-Y (p. 33). Females which have repeatedly mated with sterile males of the same strain show partial tolerance of later skin grafts from such males; 47% of the grafts surviving for five weeks or more, compared with none of a series of comparable grafts to virgin females. Similar results are obtained with females mated to normal males, when the oviducts of the females are ligated to prevent fertilization, confirming that the tolerance is a consequence of mating and not conception. However, these cases appear to be exceptional, and in most experiments mating does not appear to result in significant acceleration or retardation of the rejection of tissue grafts, whether within or between strains.

This does not, however, mean that *no* effects of mating can be detected outside the area of the uterus and its precincts. Sensitized lymphocytes, on contact with the sensitizing antigen *in vitro*, release lymphokines (p. 15) of various sorts. One has an inhibitory effect on the adherence of other leucocytes to the walls of their container. This has been exploited as the basis of a highly sensitive test for detecting a state of sensitization to a known antigen, which can be carried out on small samples of peripheral blood leucocytes.

Figure 4.10 shows the leucocyte adherence-inhibition (LAI) values obtained with the leucocytes of virgin mice, and of mice which had mated once only, four times, or seven times, with males either of the same strain, or of a different strain. A single mating suffices to raise the LAI value virtually to its maximum level in each case when leucocytes of a female are mixed with spermatozoa from a male of the same strain as the male with which she had mated. As sensitization of the circulating lymphocyte population occurs irrespective of whether the male is of the same strain as the female or not, the antigens in question cannot be strain-specific alloantigens. They could be either sperm-specific, or male-specific antigens such as H-Y.

The LAI test also reveals that sensitization of human leucocytes to seminal antigens follows coitus. Practical considerations dictated some simplification of the experimental design; no inbred strains are available for study, the women who volunteered could be divided only into three categories—virgin, non-virgin who had never conceived, and those who had previously given birth—and the spermatozoal antigens used to test the leucocytes of all women came from the same male donor. As Figure 4.11 shows, insemination in the absence of conception leads to detectable sensitization of the peripheral blood leucocytes of women, as of female mice (Dorsman *et al*., 1978).

Evidence from many sources thus indicates that the failure of the female to destroy the invading sperm before conception is not due to simple failure to recognize them as foreign: immunological recognition even of a

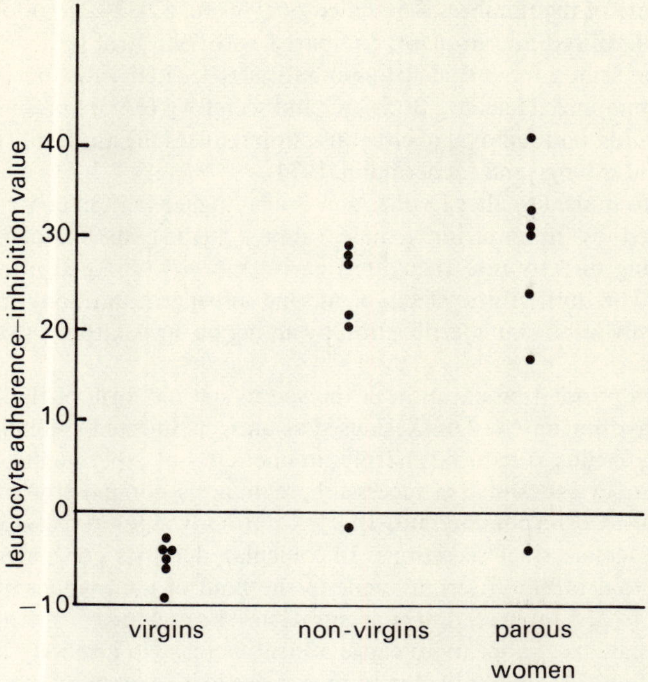

Figures 4.11 LAI values (see Figure 4.10 and p. 66) obtained with leucocytes from women who were either virgin, non-virgin and nulliparous, or parous (based on data from Dorsman *et al.*, 1978).

single mating does occur but the response which follows is not one which blocks fertilization. Nor can conception depend on failure of the efferent arm of the immune response. As already discussed (section 4.2, p. 55), soluble antibody and cells of the immune system enter the uterus and other parts of the female tract with apparent freedom. It is not clear how much of the immunity to spermatozoa depends on a cellular immune response, and how much on soluble antibody (Tinneberg *et al.*, 1980).

The secretions of the female genital tract can include antibody specifically directed against spermatozoal antigens. In rabbits, sperm-immobilizing and other antispermatozoal antibodies can be found in the blood serum following administration, by way of the vagina, of spermatozoa suspended in adjuvant. Larger concentrations of antibody result if transvaginal immunization is combined with systemic injection. Although to a lesser extent, antispermatozoal antibody is also detectable in the secretions of the uterus. When uterine fluid from an isoimmunized female is extracted and added to fresh spermatozoa *in vitro*, before using them for artificial insemination, the resulting conception rates are lowered. Taking the number of corpora lutea formed in the artificially-inseminated females as an estimate of the number of ovulated oocytes (p. 52), 24.2 % of ovulated eggs are fertilized and implant, compared with 79.2 % of the controls in which the semen for artificial insemination is treated with uterine fluid from unimmunized females. Both IgG and secretory IgA are present in the uterine fluids, and removal of either fraction reduces the antifertility effects of the fluid (Menge and Liebermann, 1974).

Embryo mortality after fertilization is also higher than in controls; this was tested by immunizing female rabbits against spermatozoa, then transferring directly into their uteri early embryos obtained from other females. The antifertility effects of uterine antisperm antibody may not consist only in blocking fertilization by acting on spermatozoa (Menge *et al.*, 1972).

When a purified preparation of the sperm-specific form of the enzyme lactic dehydrogenase (LDH-X) is used as antigen, injected systemically in adjuvant, fertility is reduced sharply; in one series of experiments, 29 % of ovulated oocytes resulted in successful pregnancies, compared with 88.3 % of controls. Corresponding anti-LDH-X antibody (IgG in class) is again found in female tract secretions. In follicular fluid, its concentration is equal to that of blood serum, while in the fluid of the oviduct it ranges from 3 % to 27 % of the serum concentration. The presence of spermatozoa in the female tract appears to cause a local increase in antibody levels in the secretion. This might be due to the release by the sperm of the enzyme hyaluronidase, known to be carried in the acrosome. Normally, its role is to help disperse the cumulus cells and so facilitate access to an oocyte; possibly it has the additional effect of locally hydrolysing the hyaluronic-acid-based intercellular cementing substances, and so increasing the permeability of the epithelia of the female genital tract to serum immunoglobulins (Goldberg, 1974; Kille and Goldberg, 1979).

In addition to the action of anti-LDH-X in preventing fertilization, there is considerable post-implantation embryonic mortality, with less than half of the embryos which implant surviving to term.

In guinea pigs, a well-defined group of sperm-specific autoantigens—S, P, T and Z—is known (pp. 34 and 41). These can be used to produce a consistent autoimmune response in male guinea pigs, rendering them infertile. Using the same antigens in a purified form, and injecting them under the same conditions, it should be equally possible to induce the formation of specific antibodies in female guinea pigs, with the injected substances now acting as isoantigens. When S, P or T antigen is injected in adjuvant (antigen Z is uncooperative; see p. 34), significant amounts of the corresponding isoantibodies appear in the blood, and in secretions of the uterine cervix and vagina. Immunized females also show positive delayed hypersensitivity reactions. The immunological effects of injecting spermatozoal antigens appear therefore to be comparable to those found in similarly immunized males (d'Almeida and Voisin, 1979).

In contrast to male guinea pigs, the females show only a trivial (and statistically insignificant) reduction in conception rates, even after two intensive series of antigen injections; in equivalent circumstances, males would be sterile. The only significant effects are an apparent delay in conception, judged by the interval between mating and subsequent birth, in females immunized against antigens T and P, and an increased incidence of fetal abnormalities (Table 4.2).

Antispermatozoal antibody can therefore be induced by sperm in the female tract, although with some difficulty, compared with alternative routes of immunization (Bell, 1969; Edwards, 1960). If artificially induced by injection elsewhere (and presumably also after natural mating) it can pass into the secretions of the female tract and make contact with spermatozoa; and yet, even in these circumstances there is scarcely any detectable effect on sperm which enter the genital tract in subsequent natural matings. How do the sperm escape?

Survival is not due to a complete lack of detectable antigens; as already discussed, the external surface of the spermatozoon is a mosaic of antigens of different types, many of which are likely to be antigenically foreign to the female. However, although sperm may not be antigenically null, a relative lowering of antigenicity would reduce the likelihood of an effective sensitization following mating, or reduce the impact of such a response if it were initiated. It has also been suggested that spermatozoal antigenicity might be reduced after insemination by enzymes of the female genital tract secretions (Katsh *et al.*, 1968).

Compared with cells such as lymphocytes, spermatozoa do carry a rather low complement of histocompatibility antigen, estimated at around 10% in the case of mouse H-2 antigen. Antisera against particular H-2 antigens have very little effect when added *in vitro* to spermatozoa carrying the antigen in question. This might be due to a reduced amount of surface H-2 antigen; however, as already discussed (p. 33), although spermatozoa have only about 10% of the H-2 antigen complement of lymphocytes, it does not follow that their surface density of antigenic sites is proportionately low. In otherwise identical test conditions, antibody against specifically spermatozoal antigens totally immobilizes sperm. The effects of these different antisera are compared in Table 4.3; it is also clear from this table that sperm-immobilizing antibody requires complement.

If the spermatozoa within an ejaculate were to any significant extent heterogeneous—by virtue of haploid expression of their genotypes, for example—this too would have a diluting effect on antigens of any one type.

Even though the importance of alloantigens may be reduced, this brings into greater prominence sperm-specific antigens. Reducing to a minimum the numbers of sperm which are at any one time permitted access to the upper reaches of the female tract might also reduce the chances of

Table 4.2 Effects on the reproduction of female guinea pigs of isoimmunization with antigens S, P and T (data from d'Almeida and Voisin, 1979)

			Delivery			
Antigen	*N*	*No. pregnant*	*N*	*Days after mating*[1]	*Litter size*[1]	*No. stillborn*
None	7	7 (100%)	7	77 ± 5.4	3.7 ± 1	0.1 ± 0.3
Spleen cells	6	6 (100%)	5	84 ± 16	3.2 ± 2	1.2 ± 2.6
S	6	5 (83%)	3	90 ± 9.2	3.3 ± 0.6	0
P	8	7 (88%)	5	117 ± 16.5	3.2 ± 1.5	0.8 ± 1.3
T	7	6 (86%)	4	85 ± 12	2.25 ± 1.5	0.25 ± 0.5

[1] Mean ± s.d.

Table 4.3 Effects of alloantisera, with and without complement, on the motility of mouse spermatozoa (data from Saji *et al.*, 1977)

	% motile sperm *Strain of mouse donating sperm*	
	C57BL/6J	C3H/He
Antiserum added		
antisperm + C′	0	0
control serum + C′	81.0	59.6
antisperm	74.5	73.3
control serum	76.0	62.5
Anti-H-2 sera		
vs. C3H/He	55.7	72.1
vs. C57BL/6J	44.9	61.6
control serum	51.9	72.6

These figures show that the major immobilizing action is that of antispermatozoal antibody, and is complement-dependent. Antisera specific to particular H-2 allotypes fails to differentiate sperm which carry those allotypes from those which do not.

sensitization; and it has been found (see above, p. 68) that sperm in the oviduct cause a local increase in antibody concentration. Rationing sperm access to the oviduct might perhaps reduce this effect. In general, only a minority of the ejaculated spermatozoa even enter the uterus in a species such as the human, and of these an almost vanishingly small number reach the upper oviduct where fertilization takes place. Most are detained in sites such as the crypts of the uterine cervix: possibly this helps to quarantine them and prevent unnecessary contact with the female's immune system.

Unwanted sperm are usually flushed away by uterine secretions, or are phagocytosed by leucocytes or uterine epithelial cells, phagocytosis being more rapid in animals which have previously been sensitized to sperm (Maruta and Moyer, 1965). There is some evidence to suggest that enzymic hydrolysis of spermatozoal antigens may also occur. These factors will all help to ensure that potential exposure of the female immune system to sensitizing spermatozoal antigen is permitted no longer than absolutely necessary.

The systemic or intrauterine injection of sperm suspensions can lead to antibody formation, and even to the secretion of antispermatozoal antibody into the female tract, without there necessarily being a significant reduction in fertility. In general, spermatozoa collected from the epididymis or vas deferens are more effective than ejaculated spermatozoa in evoking antibody formation. This might conceivably be due to a loss of surface antigen by older spermatozoa. A more likely explanation is that the antigenicity of sperm could be reduced by the contributions to the semen made by the accessory glands. Obvious candidates are the sperm coating antigens, such as the human SCA which enters seminal plasma from the seminal vesicles (p. 39). Possibly sperm-coating antigens are not antigenic in females, or only weakly so. By coating spermatozoa they could mask at least a proportion of the antigenic sites intrinsic to the sperm, so further reducing their effective antigenicity.

This conjecture is supported by the apparently greater antigenicity of human spermatozoa which are washed before injection into rabbits: possibly this removes some of the sperm-coating material so unmasking underlying sites of greater antigenicity. (In this particular case, the underlying sites may well be xenoantigenic, but the same principle would apply in the context of alloantigens or isoantigens.)

Sperm-coating antigens would not be very effective masking agents if they were themselves potent antigens. In fact human seminal lactoferrin (scaferrin) is antigenic only when injected into other species, so it does appear to fulfil this necessary requirement. Lactoferrin proper is naturally present in women. Apart from its presence in milk, it is found also in blood serum at concentrations of around 1 μg/ml, and in other contexts within the body; and although seminal lactoferrin is not totally indistinguishable from the endogenous lactoferrin of women, it is virtually indistinguishable immunologically. The major sperm-coating antigen of humans is therefore not sperm-specific. Nor does it appear to show great polymorphism, so recognition as an alloantigen is unlikely. In humans, therefore, sperm become coated with a non-antigenic material before entering the female genital tract and it seems reasonable to postulate that this coating—until it is shed—helps to prevent recognition of the coated sperm as foreign. Both initial sensitization and antibody or cell-mediated assault would therefore be impeded (Hekman and Rümke, 1969; Roberts and Boettcher, 1969).

The same may well be true of sperm-coating antigens in other species, such as the epididymal sperm-coating antigens of rabbits and hamsters (p. 39). As these were investigated by the use of antisera produced in other species, their antigenicity within individuals of the same species might be low or absent.

Masking of certain antigenic determinants of the spermatozoal surface would also be achieved by the attachment of antibody against other determinants, provided of course that the masking antibody was not itself damaging. Protection against Rhesus isoimmunization in the human fetus is to some extent provided by a similar mechanism to this (p. 148). Seemingly natural antibodies against spermatozoal antigens, as well as non-antibody spermagglutinins (one of them a β-lipoprotein), do occur in guinea pigs and rabbits, and might in some circumstances have a blocking role, or serve to accelerate the removal of sperm which could otherwise cause sensitization (Padma, 1972).

When recovered from the uterus a few hours after mating, a high proportion of rabbit spermatozoa are found to have a surface coating of immunoglobulin. Sperm collected from the male genital tract, or freshly ejaculated sperm, are generally uncoated, so the coating antibody—IgG in class—is presumably maternal in origin and attaches to sperm after copulation. The attachment of IgG appears to act as an invitation to neutrophils (p. 16) to phagocytose the coated sperm, and it has been suggested that attachment singles out senescent sperm for disposal (Hancock, 1978; Cohen and Gregson, 1978; Cohen and Tyler, 1980).

Human spermatozoa have recently been shown to carry surface receptors for the Fc region of IgG molecules, and it is possible that the affinity of spermatozoal surfaces for the Fc region may reduce the attachment of antibody molecules by their antigen-binding sites. Fc-binding would not result in complement-fixation or cytotoxicity (Allen and Bourne, 1978; Sethi and Brandis, 1980; Witkin *et al.*, 1980).

On the other hand, a proportion of the sperm recovered from the female tract after mating are uncoated by maternal IgG and remain so even when mixed with suitable IgG-containing blood serum after removal from the tract. These are the first to arrive in the uterus and form the majority of the sperm which succeed in entering the oviduct. On recovery from the oviduct and transfer into a second female, they prove, as a group, to be very fertile. Again, they fail to acquire an immunoglobulin coating (although by this time they should surely count as senescent!). Failure to adsorb antibody correlates with the apparent absence of an acrosome, so it may be that the antibody attaches to antigens on the outer membrane of the acrosome and once this is shed no further antibody-binding can occur. The female tract may therefore be selecting spermatozoa which shed their acrosomal membranes rapidly, with the remainder becoming antibody-coated and soon being phagocytosed (Taylor, 1980).

Seminal plasma contains components which are potentially antigenic and which might also provoke an immune reaction in the female. Many are in fact antigenic only in other species, and not when injected into females of the same species. Some are autoantigenic in the males which produce them (p. 45), but this does not mean they are necessarily isoantigenic when deposited in the female. Some substances, originally identified immunologically in seminal plasma, have subsequently been discovered in parts of the female as well; lactoferrin/scaferrin is one which has already been discussed. One group of investigators identified 23 antigenic components in human seminal plasma. Of these, 7 occurred also in blood serum hence could not be autoantigens peculiar to the testis, and 10 absent from serum were eventually located in other tissues and secretions of women (de Fazio and Ketchel, 1972).

Most of the seminal plasma antigens are not therefore foreign in women, and pose no immunological problem. Much of the residual evidence of antigenicity of human semen requires in any case to be reviewed in the light of the recent finding that, whereas freshly-ejaculated seminal plasma lacks antigenicity, even a few hours incubation before testing can render it antigenic as a result of bacterial modification of its constituents. Seminal plasma appears to have little or no intrinsic antigenicity (Chen and Simons, 1977). Cell-mediated immunity to human seminal plasma components has, however, been described (Marcus *et al.*, 1977).

In fact, even antibody against seminal plasma antigens, prepared in a species other than the rabbit and added to rabbit semen *in vitro* before artificial insemination, has no effect on fertility. Similarly prepared antisera against whole semen (that is, against spermatozoal as well as seminal plasma antigens) lowers fertility significantly. Even if a female did respond immunologically to seminal plasma antigens, it seems unlikely that this would make much difference to fertility (Menge and Protzman, 1967).

Apart from masking spermatozoal membrane antigens with coating substances of low antigenicity, seminal plasma may have a more active role in protecting spermatozoa from the female's immune system. Seminal plasma often has the effect of counteracting complement factors: this may serve to lessen the complement-dependent immobilization of spermatozoa which is more important in reducing fertility than sperm-agglutinating activity (p. 78), In addition, a high molecular weight component (ca. 200 000) of human seminal plasma suppresses the transformation, *in vitro*, in human lymphocytes exposed to the mitogens PHA and Con A (see p. 24), as well as their response to antigens, including the response to allogeneic cells in a mixed lymphocyte reaction (MLR). Human seminal plasma components act also on mouse lymphocytes, preventing the induction of a cell-mediated cytotoxic response, although not the expression of one previously induced. The response to PHA, and both T-cell-dependent and T-cell-independent responses to specific antigens, are also suppressed (Lord *et al.*, 1977; Marcus and Hess, 1980; Stites and Erickson, 1975; Tarter and Anderson, 1980).

At least two different seminal components are probably involved, of which one is apparently a protein. Other effects may be caused by seminal prostaglandins. Certain of the prostaglandins—a family of unsaturated fatty acids with a variety of potent physiological effects—are known to have modulatory effects on immune responses, including suppression. A range of prostaglandins is added to human semen by the seminal vesicles (not, despite the name, by the prostate).

Although these various immunosuppressive actions have been demonstrated only *in vitro* and their relevance to the circumstances of natural insemination has not been established, seminal plasma does appear to have some ability to manipulate immune responses in the female genital tract, to the potential advantage of spermatozoa.

After copulation, at most a tiny minority of the sperm in the ejaculate achieve fertilization. The rest soon die and are eliminated in various ways. In several species of rodent, sperm heads have also been found within epithelial cells of the oviduct. Uterine cells may also take in spermatozoa while regenerating after injury. This 'somatic fertilization' has led to the suggestion that some spermatozoal DNA, entering somatic cells in a relatively intact state, might express its genetic information, and that this supposed expression might result in the expression of antigens not otherwise present in the female. If so, then transformed maternal cells could act as amplifiers of the antigenic constitution of the sperm, or as a means of prolonging the exposure of the female to spermatozoal antigens after the demise of the actual spermatozoa. This might conceivably help to explain some of the apparent eccentricities of the response of the female tract to spermatozoa (Reid, 1965; Reid and Blackwell, 1967).

Experiments *in vitro* confirm the possibility of relatively intact DNA being taken up by somatic cells, such as lymph node cells and peritoneal macrophages; some is even found within the nuclei of a small proportion of macrophages. No evidence has ever been presented of any activity of this DNA. In fact the only evidence of any genetic implications of spermatozoal entry into somatic cells of the female genital tract comes from the finding that cells of the HeLa strain, of blood group O, can acquire group A antigenicity if cultured with spermatozoa of group A. HeLa cells were among the first human cells to be successfully established as an *in vitro* culture line; as they stem from a (fatal) cervical tumour, and have altered considerably in the course of several decades of laboratory culture, they are probably rather less representative of normal human cells than a Pekinese dog is of its original wild ancestor. Any immunological significance of 'somatic fertilization' has yet to be established; at present it remains merely an intriguing possibility (Beer and Billingham, 1976).

Survival by at least some spermatozoa of the immunological hazards of the female genital tract is therefore not yet adequately understood, despite a considerable research effort. Many factors probably contribute, although some of the suggested mechanisms are probably of, at most, marginal importance (Hancock, 1978; Rümke, 1980).

4.4 Immunology and infertility in women

No biological system is totally reliable; whatever the mechanisms which normally protect spermatozoa against the female's immune system, one would expect occasional breakdown, with evidence of sterility, or impaired fertility, attributable to an immune response. Infertility in woman might therefore correlate to some extent with the presence of antibody against semen, or with a cell-mediated immune reaction against spermatozoa.

There have been few investigations of the prevalence of cell-mediated immunity to spermatozoa, and its relation to infertility. As an indicator of a state of cell-mediated immunity, one study used the release of the lymphokine MIF (macrophage migration inhibition factor; see p. 24) when suspected lymphocytes were exposed, *in vitro*, to spermatozoal antigen. The antigen for the test was obtained from washed spermatozoa, pooled from several donors, so that a significant MIF response would indicate sensitization to antigens characteristic of sperm, rather than to the husband's alloantigens.

Using this criterion, it was found that in a group of infertile women where no other cause of infertility was known (that is, excluding such causes as tubal obstruction and endocrine disorders) the incidence of positive MIF responses was 74%; this compared with 62% of fertile women, and 0% of virgins (girls between 9 and 16 years old). All of a further sample of infertile women with circulating humoral antibody to spermatozoa—spermagglutinating and spermimmobilizing—showed a positive cell-mediated response to spermatozoal antigens. The difference between the fertile and infertile women was not statistically significant, so any link between infertility and cell-mediated immunity must be a tenuous one. These results accord with those of the comparable study discussed on p. 67, which used a different lymphokine assay to detect sensitization to spermatozoa, and did not take fertility into account (Mettler and Schirwani, 1975).

The presence of soluble antibody against spermatozoa is easier to assess than cell-mediated immunity, so most studies have concentrated on this. For equally practical reasons, published results relate almost exclusively to serum concentrations of antibody. Although the antibody concentrations faced by spermatozoa within the female genital tract are probably related to serum concentrations, direct assessment of antispermatozoal antibody in oviduct or uterine fluid, or in the cervical mucus, would be preferable.

Detection of sperm-agglutinating antibody is, in principle, easy to achieve. A sample of serum to be tested, added to live sperm in some form of culture vessel, will cause agglutination if agglutinin is present, and not otherwise. In practice, the technique is less straightforward, and much energy has been expended in discussing the niceties, and relative merits and demerits, of different modifications of this basic approach.

The serum dilution used is critical. At high serum concentrations, a degree of non-specific agglutination occurs. Part at least of this may be due to the β-lipoprotein agglutinin mentioned on p. 46, which is found in the serum of women as well as in that of men. The optimum concentration of serum is therefore one which is sufficiently dilute to eliminate such non-specific effects, without lowering specific spermagglutinating antibody concentrations to levels where detection becomes unreliable. This is largely a matter of judgement, with different investigators adopting different dilutions.

Assessment of agglutination also depends on technicalities. Two main procedures are commonly in use. One, due originally to Franklin and Dukes, but the basis of many subsequent improvements, depends on direct observation of spermatozoa under a microscope, and recording aggregation or the failure to aggregate. When aggregation does occur, it is usually of the head-to-head variety, possibly suggesting the predominance of IgM (p. 46). Clearly there is considerable scope for subjectivity in this method, and in the earlier uses of this test, among other shortcomings, no attempts were made to standardize the criteria being used by different workers. This led to widely divergent estimates of the incidence of antispermatozoal antibodies.

The principal alternative technique, due initially to Kibrick, uses spermatozoa suspended in a 5% gelatin solution with different serum dilutions, and the estimate of agglutination from the macroscopic appearance of the suspension. Here the agglutination is largely tail-to-tail, indicating sensitivity to a different antigen–antibody combination from that which forms the basis of the Franklin–Dukes test (Beer and Neaves, 1978; Shulman, 1972, 1978).

Table 4.4 gives the results of one survey, where the two techniques were applied to the same serum samples. They appear to give gratifyingly similar results. This apparent agreement is, however, largely spurious. Whereas 4.6% of samples were positive according to both methods, and 71.4% negative by both, a total of 24.0% of samples gave discordant results, being positive by one method and negative by the other.

Perhaps the most unexpected finding, shown in this table, is the high proportion of pregnant women who were recorded as having spermagglutinating antibody in their blood serum. This is a point on which different surveys tend to disagree. Table 4.5 illustrates for comparison a selection of surveys which used variants of the Franklin–Dukes technique. Differences between investigators apart (and these are considerable) it does appear that in general infertile women (or at least the female partners in infertile marriages), when no other cause of infertility is apparent, have an increased incidence of serum spermagglutinating antibody. In pregnant women, the incidence may be high or low, depending on the investi-

Table 4.4 Comparison of Franklin–Dukes and Kibrick tests for spermagglutinating antibody in women, and incidence of sperm-immobilizing antibody. All figures quoted are % of sample (data from Isojima *et al.*, 1974).

Status of women	*Franklin–Dukes*	*Kibrick*	*Sperm-immobilizing Ab*
Infertile: cause unknown	40	35	18
known cause	25	25	1
Control: pregnant women	45	38	0
unmarried women (undetermined fertility)	9	11	0

Table 4.5 Incidence of spermagglutinating antibody in the serum of women in various reproductive categories, judged by the Franklin–Dukes method, as employed by 3 different investigating groups. A—Franklin and Dukes (1964) *Am. J. Obstet. Gynecol.* **98**: 6; B—Kolodny *et al.* (1971) *Obstet. Gynecol.* **38**: 576; C—Schwimmer *et al.* (1967) *Fertil. Steril.* **18**: 167.

Status of women	*A*	*B*	*C*
Infertile: cause unknown	79	14	42
known cause	10		20
Fertile women	12[1]	0[1]	3
Pregnant women	—	—	34
Unknown fertility	4	—	20
Prostitutes	—	17	73
Nuns	—	4	—

[1] Includes pregnant and parous women.

gator, while prostitutes—experiencing more frequent exposure to seminal antigens—tend to show a relatively high incidence. This may possibly be related to their relatively low fertility; a point adumbrated, curiously enough, by Charles Darwin long before the subject of immunology was thought of.

Sperm-immobilizing antibody may also be present, and is generally assessed by measuring the time taken for the mobility of a sample of spermatozoa, under closely-defined conditions, to fall to a specified level compared with a control sample in the absence of serum. Sperm-immobilization is complement-dependent and closely associated with cytotoxicity, although it is not yet clear whether immobilization and cytotoxicity are different aspects of the action of the same antibody.

Some results of sperm-immobilizing tests are also given in Table 4.4 for comparison with the results of agglutination tests. It is apparent that infertility is more closely related to the presence of immobilizing antibody than to spermagglutinin. Table 4.6, showing the occurrence of pregnancy following detection of antispermatozoal antibody, confirms this; particularly in the case of couples who had been infertile for three or more years previously.

Table 4.6 History of conception subsequent to the determination of spermagglutinating or sperm-immobilizing serum antibody—women with known cause of infertility excluded from analysis (data from Jones, 1980)

		% pregnancy reported among women who were	
		(*a*) *described as infertile*	(*b*) *had been so for 3 or more years*
Agglutinating Ab	absent	53.6	44.4
	present	41.4	37.5
Immobilizing Ab	absent	52.8	45.4
	present	34.8	13.3

Establishing the presence of antispermatozoal antibody in the serum of infertile women does not, of course, shed much light on how or where antibody affects spermatozoa after insemination. The first major obstacle to the passage of sperm is the uterine cervix and, as antibody is known to be present in the cervical mucus (p. 57), this is likely to be one point at which sperm are impeded. From the point of view of the spermatozoon, the situation is probably much the same whether the antibody is added to the semen before ejaculation (p. 40) or is present in the cervical mucus as a result of secretion by the female genital tract itself.

Sperm from infertile men with significant levels of seminal spermagglutinating autoantibody show poor ability to penetrate a sample of normal cervical mucus *in vitro*, while normal sperm do not readily penetrate cervical mucus which contains antispermatozoal antibody of female origin. Normal sperm almost completely lose their capacity to penetrate normal mucus if they are first mixed with blood serum containing spermagglutinating antibody; the progressive forwards motion of swimming sperm, an important factor in their passage through mucus, is replaced by a rapid and jerky pattern of movement as soon as contact is made with the edge of a patch of mucus. This effect is not dependent on the presence of complement (unlike the action of sperm-immobilizing antibody) and appears to be due to IgA rather than IgG (Fjällbrant, 1968*b*; Kremer *et al.*, 1978).

Figure 4.12 illustrates an interpretation of this phenomenon. If antibody is present in the seminal plasma, it coats the spermatozoon from head to tail. On coming into contact with cervical mucus, the coating antibody adheres to the glycoprotein micelles of the mucus, so anchoring the spermatozoon to the mucus surface. It is trapped like a fly on flypaper, and is no longer capable of the forward movement necessary for penetration of the mucus. If the antibody is present in the cervical mucus rather than in the seminal plasma, it tends to trap only the head of the spermatozoon. The effect on sperm movement is the same.

There are therefore clear correlations between the spontaneous presence of antispermatozoal antibodies and infertility in women (and in men). The chief purpose of investigating this correlation is not, however, to develop a technique sufficiently sophisticated to discriminate accurately between fertile and infertile women; failure to conceive is obviously the criterion which matters. What immunological investigations might supply is the identification of a contributory cause in specific cases, and the suggestion of possible approaches to therapy.

A primary aim of therapeutic measures has been to lower the levels of antispermatozoal antibody. A prolonged period of condom use, or of abstinence from sexual intercourse, has sometimes been followed by a decline in circulating antibody. In a few cases, pregnancy has also ensued; however, analysis of the data suggests that the likelihood of pregnancy is

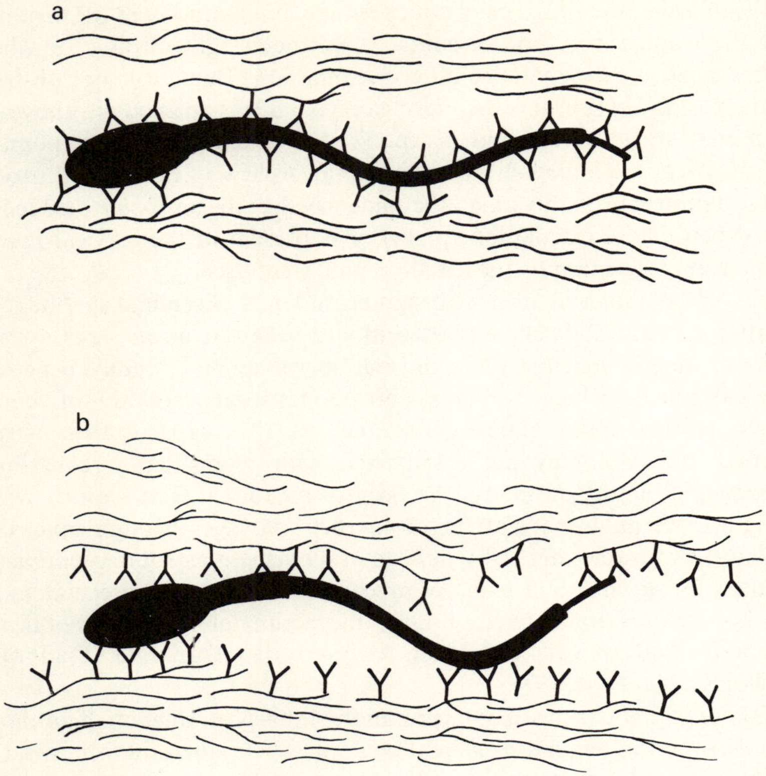

Figure 4.12 Interpretation of the 'shaking' and failure to penetrate cervical mucus by human spermatozoa in the presence of antispermatozoal antibody. (A) represents the situation where antibody is present within the seminal plasma; in (B) antispermatozoal antibody is within the cervical secretions (redrawn from Kremer, 1978).

somewhat less than if no therapeutic measures had been attempted. Immunosuppressive drugs, such as corticosteroids, have also been tried, but the results were not impressive enough to outweigh the hazards of general immunosuppression. Given that the uterine cervix appears to be the major barrier that sperm must overcome, an alternative approach is to bypass the barrier by introducing the husband's semen directly into the uterus. This appears to be a promising technique; in one case four pregnancies resulted from 20 attempts, and in another four out of seven women became pregnant as a result of intrauterine insemination.

So far, antibodies against spermatozoal antigens have been considered. As already pointed out (p. 36), certain seminal plasma constituents are also antigenic and potentially capable of causing immune responses, although many are at most weakly antigenic in members of the same species. Two cases are on record of violent allergic responses to seminal plasma components, in the absence of any detectable antispermatozoal antibody. In one case, 'within 5 minutes after her husband's ejaculation, itchy, swollen eyes, nasal obstruction and sneezing developed. ... she also noticed swelling and itching of the perivaginal area, diffuse urticaria (rash) and a sensation of swelling in her throat. These symptoms gradually subsided over a period of 4 days.' (Levine *et al.*, 1973.)

In another case, the effects were even more distressing, and culminated in cardiovascular collapse and loss of consciousness (Halpern *et al.*, 1967). Both allergens were identified as proteins, the latter as a glycoprotein rich in sialic acid. Because of the widespread nature of the symptoms, it seems probable that allergen molecules managed to leave the uterus, enter the circulation, and pass round the body in an antigenically unaltered state. The antibody in question is presumably IgE. A curious case of coital sensitization to dog (Alsatian) semen probably also involved IgE, but in this case there is no information on the nature of the allergen (Holden and Sherline, 1973).

Cellular immunity to spermatozoa may also develop, and correlates with infertility in women: in one study, the lymphocytes of 18 % of a group of infertile women showed MIF release when cultured together with sperm (p. 24); the corresponding figure for normally fertile women was 0% (Shulman *et al.*, 1977).

Women may therefore on occasion become sensitized to components of semen, and this may adversely affect fertility. This is naturally distressing to a couple who wish to conceive, but the relative infrequency of infertility of immune origin, and the extreme rarity of allergic responses to semen, throw into relief the remarkable, and still largely unexplained, routine failure of the female immune system to respond in a hostile way to insemination.

Finally, autoimmunity can develop against organ-specific antigens of the female reproductive system and may, as in the male, be associated with infertility. Antibody against granulosa or thecal cells of the ovary is known, and may lead to endocrine failure and sterility. Frequently such antibodies react also with the adrenal cortex, hence the autoantigens in question are presumably characteristic of steroidogenic tissue in general (Irvine and Barnes, 1974; Jones, 1980). Antibody formation against HCG or LH has also been reported (Wass *et al.*, 1975). The occurrence of autoantibody against zona pellucida is associated with infertility, although women of normal fertility may also show such antibody (Aitken *et al.*, 1981; Sacco, 1977; Shivers and Dunbar, 1977). Rarely, infection with mumps virus may cause an autoimmune reaction in the ovary similar to the autoimmune orchitis which commonly afflicts men infected with this disease (p. 48). A number of antigens characteristic of various parts of the female reproductive system have been identified (Stone *et al.*, 1980; Vallotton and Forbes, 1966), thus it is possible that further autoimmune reactions will be discovered in women. As in men, however, it may be that infertility and autoantibody formation are both effects of the same initial cause; a correlation between the two does not prove that autoantibody formation causes infertility (Jones, 1980; Scott and Jones, 1976).

CHAPTER FIVE

IMMUNOLOGICAL IMPLICATIONS OF PREGNANCY

5.1 Relations between mother and fetus

Probably the only way to make the complexities of the immunological relations between semen and the female genital tract seem, by comparison, relatively simple and straightforward is to consider the nature of the immunological relationship between mother and conceptus. This was touched on briefly in Chapter 1, and must now be considered in more detail. As always in reproductive biology, a disproportionate amount of the available information comes from a disproportionately small number of species, notably rats, mice, and humans. Considering that mammals range in size from shrews (a few grams) to the blue whale (150 tons), and in gestation length from 16 days (hamster) to 645 (Indian elephant), it would be most remarkable to find a high degree of uniformity in the detail of maternal–fetal relations. What is found to be true of mouse or human may well not prove to apply, in more than the most general way, to other mammalian species (Austin and Short, 1972*d*).

In the mouse, as in most mammals, fertilization takes place in the upper reaches of the oviduct. By about $3\frac{1}{2}$ days after mating, the embryo has passed through the cleavage stages of development, shed its surrounding cumulus cells and zona pellucida, and entered the uterus as a blastocyst of some 60 or so cells, rather less than 0.1 mm in diameter. At this stage, it is clearly divisible into the *inner cell mass* (from which comes the embryo proper, as well as certain of the extraembryonic membranes) and the surrounding layer of *trophectoderm* (Figure 4.1). Implantation starts on day 4. Firstly, the trophectoderm opposite to the inner cell mass, the *mural trophoblast*, begins to extend and alter in appearance. Transformed mural trophoblast cells, enlarged and with a DNA content increasing progressively to several hundred times that of normal diploid cells, are known as the primary giant cells. These may be as much as 0.5 mm in diameter, with a nucleus of 0.1 mm even visible to the naked eye (Austin and Short, 1972*a*, 1972*b*; Balls and Wild, 1975; Hogarth, 1978).

The trophoblast which lies over the inner cell mass, or *polar trophoblast*, then grows out into the decidualized endometrium as a cone-shaped structure known as the *ectoplacental cone*. This will give rise to the bulk of the placenta. From its margins arise secondary giant cells, which spread away from the ectoplacental cone and eventually form a layer around the entire conceptus. The remainder of the ectoplacental cone trophoblast continues to proliferate and invade the endometrium. The walls of the maternal capillaries it encounters are disrupted, so that it comes to be bathed directly in maternal blood, in the form of a network of sinuses within the trophoblast mass. After further differentiation, the trophoblast outgrowth forms three major regions: the *giant cell layer* (which diminishes in late pregnancy and disappears before birth), and two regions where the trophoblast forms a syncytium, the *spongiotrophoblast* and the *labyrinthine trophoblast*, both—as their names suggest—riddled with maternal blood spaces. Figure 5.1 illustrates trophoblast outgrowth during implantation, and Figure 5.2 the descent of the different cell types involved (Balls and Wild, 1975; Gardner, 1975).

Trophoblast is homologous with the chorion and therefore represents one of the major extraembryonic membranes of the higher vertebrates. Penetration of the chorionic trophoblast by another of the extraembryonic membranes, the richly vascular *allantois*, endows it with a copious blood supply and completes the establishment of the definitive chorioallantoic

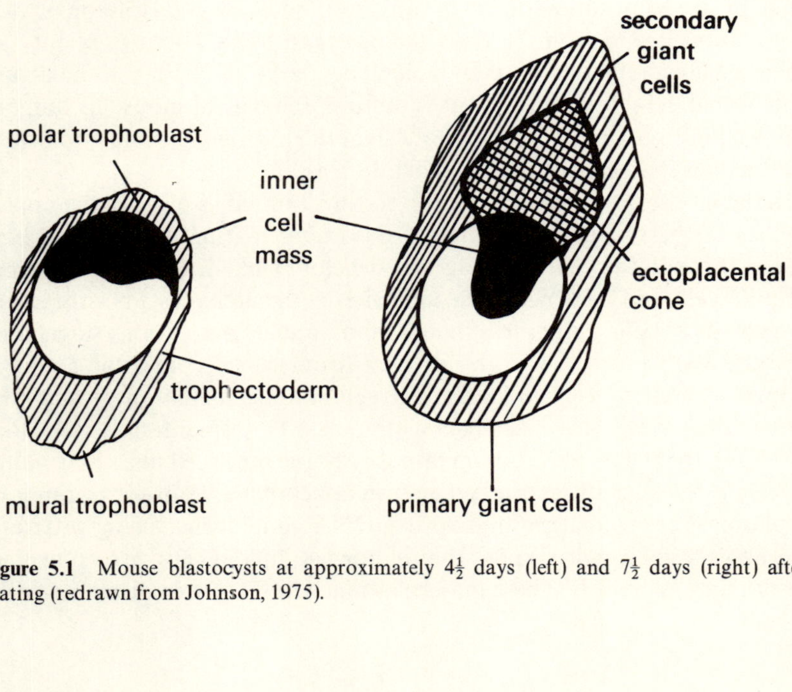

Figure 5.1 Mouse blastocysts at approximately $4\frac{1}{2}$ days (left) and $7\frac{1}{2}$ days (right) after mating (redrawn from Johnson, 1975).

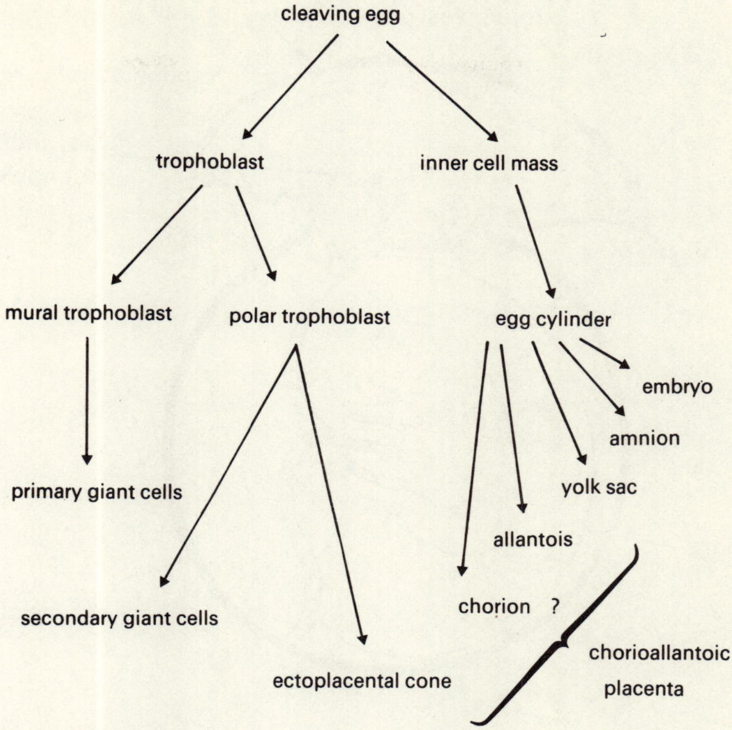

Figure 5.2 Constituent tissues of the conceptus and their origins (redrawn and adapted from Gardner, 1975).

placenta. Because of the formation of the blood sinuses within the trophoblast, the maternal and fetal circulations, at their closest, are separated only by the endothelial walls of the fetal capillaries and a thin layer of trophoblast.

The non-placental regions of the chorion also lie in direct contact with maternal tissues, while additional contact occurs between the fetal yolk sac and the uterus. In mice, the yolk sac is an extensive organ, which comes virtually to surround the embryo except on its placental side. By the breakdown of the intervening membrane in the second half of pregnancy, the yolk sac endoderm directly abuts the endometrium (Figure 5.3).

During implantation of the human embryo, the blastocyst sinks completely into the endometrium. Trophoblast grows out initially in the form of a thin cellular layer, the cytotrophoblast, and a syncytial outer layer (Figure 5.4a). The outer surface becomes extended into numerous villous projections. Columns of the underlying cytotrophoblast erupt through the syncytiotrophoblast at the apices of these villi, spread laterally, and fuse to form a shell round the embryo. The spaces between this outer shell and the

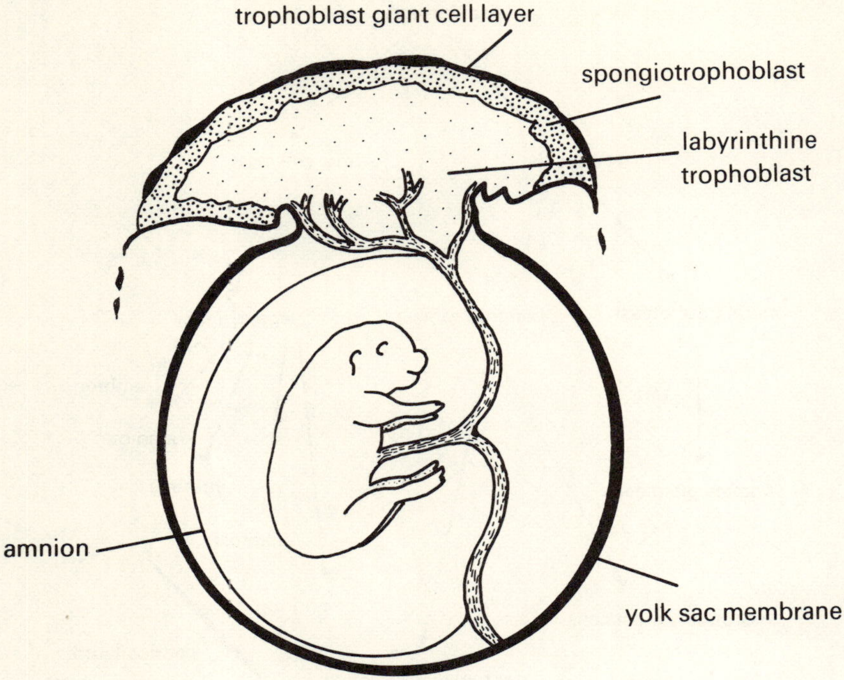

Figure 5.3 Mouse embryo during second half of pregnancy, showing the arrangement of the fetal membranes (redrawn from Billington, 1975).

inner layers of trophoblast are subdivided by further villous outgrowths, and the passage of maternal blood through the labyrinth of intervillous spaces, as well as through interconnecting channels within the trophoblast, ensures extensive contact between the syncytiotrophoblast of the villi and the surrounding maternal blood.

On the fetal side, vascularization of the placenta is not, as in the mouse, by means of the allantois but by vessels penetrating directly from the fetus along the connecting body stalk and into the chorionic placenta. Later in pregnancy, the cytotrophoblast becomes discontinuous and the syncytiotrophoblast and the endothelium of the fetal capillaries alone stand between the two blood systems. Also in contrast to the arrangement in mice, the human yolk sac is small (Figure 5.4) and does not come into direct contact with maternal tissues. It has been reported that during early implantation trophoblast and decidual cells actually fuse (as is the case in the implantation of the rabbit embryo) but this has not been confirmed (Steven, 1975; Tekelioğlu-Uysal *et al.*, 1975).

A feature of human pregnancies is that considerable numbers of syncytiotrophoblast cells break away from the placental villi and are swept

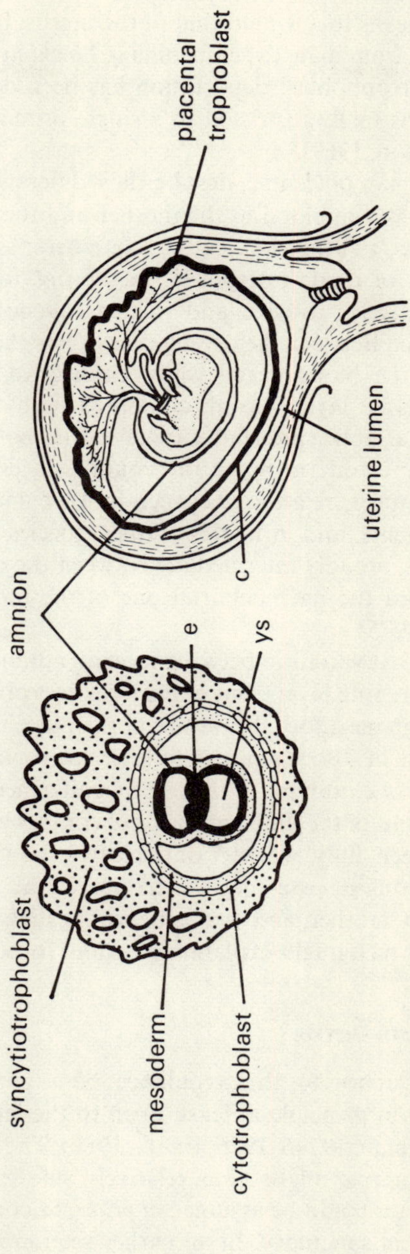

Figure 5.4 Development of the human fetus: (left) early implanting embryo, (right) established pregnancy. (ys = yolk sac; e = embryo; c = chorion) (redrawn from Billington, 1975).

off in the maternal blood. Many of these cells are probably destroyed, but the remainder survive and are filtered out by the next capillary bed they meet, in the lungs. Here they may survive for some time, and may disrupt surrounding lung tissue, as they would that of the uterus, but little lasting damage is done. Apart from man, the chinchilla (*Chinchilla lanigera*) is the only species in which trophoblast deportation has been described, but it seems unlikely that it is really restricted to a single primate and a single rodent species (Billington, 1969).

Both mouse and human placentae, despite their differences, are classed as haemochorial; that is, the blood of the mother and the chorion of the fetus lie in direct contact. In other species, the relationship is a less intimate one, with more layers of tissue interposed. In the pig, for instance, the trophoblast is not actively invasive and in consequence the maternal capillaries remain intact. Between the two blood systems there are, on each side, endothelium of the blood vessel walls, connective tissue, and an epithelial layer: six tissue layers in all, compared with the two of the haemochorial placenta. Intraepithelial capillaries in both uterine epithelium and chorionic ectoderm make the contact, in effect, somewhat closer; in general, however, relations between mother and fetus are less intimate than in rodents and primates. Other species have different placental relationships, broadly intermediate between the epitheliochorial placenta of the pig and the haemochorial one of the mouse or human (Perry, 1981; Steven, 1975).

There is therefore great variation between species, not only with respect to the number of intervening layers, but of the thickness of these layers. At different stages of pregnancy, too, the nature of the placental barrier may alter; the development of discontinuities in the cytotrophoblast layer of the human placenta is one example of this. The significance of the varying types of trophoblast, and of the different forms of contact between mother and fetus, has not been fully worked out; clearly, however, these are important considerations in assessing the nature of the immunological interaction between a mother and her genetically alien fetus, and the possible mechanisms which might explain her curious forbearance.

5.2 How might the fetus survive?

Many different approaches to the avoidance of a harmful maternal immune response are, in principle at least, open to the fetus (Beer, 1979, 1980; Beer and Billingham, 1974*b*, 1976; Faulk, 1980). The uterus, particularly a decidualized uterus, might be a relatively safe site in which an allograft—such as a fetus could be assumed, *a priori*, to constitute—might be given immunological sanctuary. In an earlier section (pp. 59 and 62) this was shown not to be the case for seminal antigens or artificially introduced cell suspensions or other allografts; nevertheless a pregnant uterus might confer some degree of immunological privilege on its contents.

A female might while pregnant be less capable than at other times of detecting and rejecting foreign antigens. A generalized depression of the immune system might be caused by the high levels of certain hormones which prevail during pregnancy; or by certain other substances which are characteristically present during pregnancy. Alternatively, as the placenta is a site of hormone synthesis (p. 54), the local application of hormones (or other substances) might locally impede immune recognition or destruction of the fetus.

So far, the fetus—or, strictly speaking, the conceptus: all of the cellular products of conception, whether fetal or extraembryonic—has been assumed to be antigenic. Possibly it might defer expression of histocompatibility antigens of paternal origin until it is too late for a maternal immune response to do any damage. Alternatively, a normally antigenic fetus might shelter behind some form of barrier which prevented access either of fetal antigen to the maternal immune system, or of antibody and sensitized lymphocytes in the reverse direction. A requirement of such a barrier is that it should itself be either incapable of sensitizing the mother, or not susceptible to immunological attack. As trophoblast lies between maternal and fetal tissues, it is an obvious candidate for scrutiny in this context.

Finally, fetal survival might depend, not on the failure of an immune response to develop, but on the nature of the response. The fetal parasite might be able to manipulate its maternal host in such a way as to induce acceptance instead of rejection.

In the event, none of these possible mechanisms, on its own, supplies a single, simple explanation of fetal survival; all appear to some extent to be relevant in an enormously complex situation. They will be discussed in turn.

5.3 Immune response of the oviduct and uterus to embryos

Before implanting in the uterus, fertilized eggs spend a number of days in the oviduct. While there they pass through the cleavage stages of development and become blastocysts. Very little is known of their immunological position, in relation to the surrounding oviduct, during this period. Assuming that cleavage and early blastocyst stages actually carry histocompatibility antigens—a complex question discussed further in section 5.5—this raises the issue of the immune competence of the oviduct.

It is not known whether an immune response can be initiated from within the oviduct, although soluble antibody is found in oviductal fluids (p. 56). The zona pellucida is permeable to immunoglobulins, including the large molecules of IgM as well as IgG. By the mmediate post-implantation period, significant amounts of antibody can be detected in various parts of the mouse embryo, particularly among trophoblast and yolk sac membrane cells. This antibody does not appear to be harmful, and may indeed be beneficial to the embryo, or even vital for its survival (Bernard, 1977; Bernard *et al.*, 1977; Sellens and Jenkinson, 1975).

There is no direct evidence of destructive immune reactions against mammalian embryos before implantation; such evidence would be difficult to obtain. It must, however, be remembered that a high proportion of zygotes probably dies during this period: between 10% and 25% in certain strains of mouse (Boshier, 1968), and at least as high a proportion (by inference) in humans. It is possible that deleterious immune reactions by the mother contribute to this early mortality.

More is known of the situation once the embryo enters the uterus. A skin allograft to the rat uterus is, as discussed previously, rejected as one to another part of the body. Only if the uterus is in a receptive state during early pregnancy or pseudopregnancy, and decidualization has been induced, will there be any prolongation of graft survival. Evidently some degree of protection is conferred on its contents by a decidualized uterus. An element in this protection appears to be the restriction of afferent lymphatic pathways so that access of graft antigens to the maternal immune system is impeded. Alternatively, patrolling lymphocytes which encounter foreign antigen by coming into contact with the graft might find it difficult to return to a local lymph node: this, too, would hinder the initiation of an effective immune response. The significance of an afferent block is shown by bypassing the impediment: if the recipient of an allograft has previously been hyperimmunized against graft alloantigens, the efferent arm of the immune system has no difficulty in obtaining access to the uterine lumen and destroying a skin allograft there.

A pregnant female rat intensively immunized against paternal alloantigens, in sharp contrast, conceives and successfully gestates her litter despite her state of hyperimmunity. Fetal survival cannot therefore depend solely on an afferent block to sensitization, since any such block has in this case been bypassed. The defence that the fetus possesses in this case, and the skin allograft lacks, may be the intervening layer of trophoblast. The quarantining properties of trophoblast are discussed later, in sections 5.6 and 6.5 (Beer and Billingham, 1974*a*, 1976).

Further evidence against any significant role of restriction of afferent lymphatic drainage from decidual tissue is supplied by the considerable hypertrophy which occurs in the para-aortic lymph nodes which drain the uterus of rats, mice and hamsters in response to allopregnancy. This is similar in magnitude to (although rather smaller than) the hypertrophy which follows an intrauterine skin allograft. The effect of syngeneic pregnancy is statistically insignificant, indicating that it is alloantigens of the conceptus, rather than antigens characteristic of embryonic or tropho-

Table 5.1 Weight of lymph nodes draining the uterus in syngeneic and allogeneic pregnancies in rats, mice and hamsters. Weights quoted in mg ± s.d. Note that in rats Fischer and Lewis strains are identical at the major histocompatibility locus AgB and differ only with respect to minor loci; strain DA differs from both at the AgB locus (data from Beer and Billingham, 1974).

	Genetic strain of		
Species	*Female*	*Male*	*Mean weight of draining lymph nodes*
Rat	Fischer	—[1]	7.04 ± 3.6
	Fischer	Fischer	7.75 ± 3.47
	Fischer	Lewis	22.38 ± 3.9
	Fischer	DA	29.46 ± 9.33
Mouse	C57	C57	1.8 ± 0.8
	C57	A	4.7 ± 1.7
Hamster	MHA	—[1]	2.73 ± 0.78
	MHA	MHA	3.3 ± 1.3
	MHA	CB	9.2 ± 4.5

[1] Non-pregnant controls.

blast cells, which act as the stimulus (Table 5.1; compare Figs. 4.7 and 4.8 and Table 4.1)—see Beer and Billingham, 1974*a*, 1976. Curiously, the corresponding regional lymph nodes of women show the opposite tendency and decrease in size and cell content during pregnancy (Nelson and Hall, 1964).

Apart from establishing that the uterus does not preserve its fetal contents by isolating fetal antigens, these experiments obviously demonstrate that some part of the conceptus does in fact carry histocompatibility antigens. The possibility that fetal survival is permitted because of a simple failure to develop paternal alloantigens can therefore be dismissed.

Local sensitization of the uterus to antigen is also possible, as indicated by the 'recall flare' reaction (p. 62). An unexpected consequence of this is that local presensitization of the rat uterus, far from jeopardizing subsequent pregnancy, actually has a beneficial effect. The mean number of embryos developing in a sensitized uterine horn is increased, compared with the number in an unimmunized contralateral horn (Table 5.2). Whether this significantly affects reproductive performance in normal circumstances is not clear (Beer and Billingham, 1974*a*, 1976).

Conceptuses within a uterus are therefore treated in some respects differently from allografts of other tissues there. The contribution of the uterus to the survival of allogeneic embryos can also be investigated by comparing development within the uterus with development elsewhere, in an ectopic site within the body. Occasionally, human blastocysts escape the confines of the oviduct and implant within the peritoneal cavity in positions such as the exterior wall of the large intestine. Many such ectopic

Table 5.2 Effects of prior immunization of one horn of the rat uterus on the subsequent reproductive performance of that horn.

Female rats of the Fischer strain (FI) were recipients of a sensitizing skin allograft from donors of either the Lewis (LE) or DA strains, placed in the right horn of the uterus. Three weeks later, intense local hypersensitivity reactions could consistently be evoked. The females were then mated to rats of the allograft donor strain or, as a control, to rats of their own strain. At 18 days of pregnancy the number of fetuses in each horn was counted. Note that Fischer and Lewis rats are identical at the MHC locus, and both differ from DA rats (data from Beer and Billingham, 1974).

Recipient female	*Sensitizing strain*	*Paternal strain*	*Mean number of embryos in*	
			Right horn	*Left horn*
FI	LE	LE	4.8	3.6**
FI	LE	FI	4.5	4.7
FI	DA	DA	5.1	3.2*
FI	DA	FI	3.3	3.2

* $P < 0.05$.
** $P < 0.01$.

pregnancies die (often without their existence being suspected), although there is no evidence that recognition of their alloantigenic status, rather than nutritional or mechanical factors, determines this. A few, however, survive for long periods, even to term; in a very few cases a live baby has been surgically delivered. The uterus is not indispensable.

David Kirby, among others, achieved similar results artificially, and showed that blastocysts, recovered from the uteri of pregnant mice and transferred to ectopic sites, could develop there. Beneath the capsule of the kidney or testis were two particularly suitable sites; in fact trophoblast outgrowth in such ectopic sites was enthusiastic and unrestrained, and the results correspondingly devastating. Apart from more rapid blastocyst outgrowth, the overall success rate of blastocyst transfers to, for example, the testis was higher (almost 100%) than that of similar transfers to a hormonally prepared uterus (typically around 40%). Mouse blastocysts could also be transferred within the testis capsule of male rats; even with the wrong site, the wrong sex, and the wrong species, considerable trophoblast outgrowth and embryonic development took place (Kirby, 1965, 1968, 1970).

If the recipient of a mouse blastocyst was an allogeneic mouse which had previously been hyperimmunized against the alloantigens of the blastocyst, growth was better in the uterus than in an ectopic site such as the kidney (Searle *et al.*, 1974). In less extreme circumstances, when the recipient has not been hyperimmunized, there is little to choose between the two organs. The uterus, no doubt in many respects an excellent place in which to have a pregnancy, is from an immunological point of view not the only place possible. The uterine contribution to immunological survival of the fetus appears slight.

5.4 Depression of the mother's immune response

One possible explanation for fetal survival is that the female, by virtue of her pregnancy, becomes less able to mount an immune response to any antigenic stimulus with which she is faced. This could prevent immunological recognition of her fetuses, or delay a response, possibly until after birth, when it would no longer affect the offspring. Alternatively, an attenuated response might give a fetus the chance to impose a state of specific tolerance on its mother.

If generalized depression of the immune system was a consequence of pregnancy, then it would apply to allografts other than fetal ones. Pregnant rabbits do, on average, take slightly longer than non-pregnant ones to reject allografts of skin; the extent of the reprieve varies, but at most amounts to about a doubling of survival time. Pregnancy does not

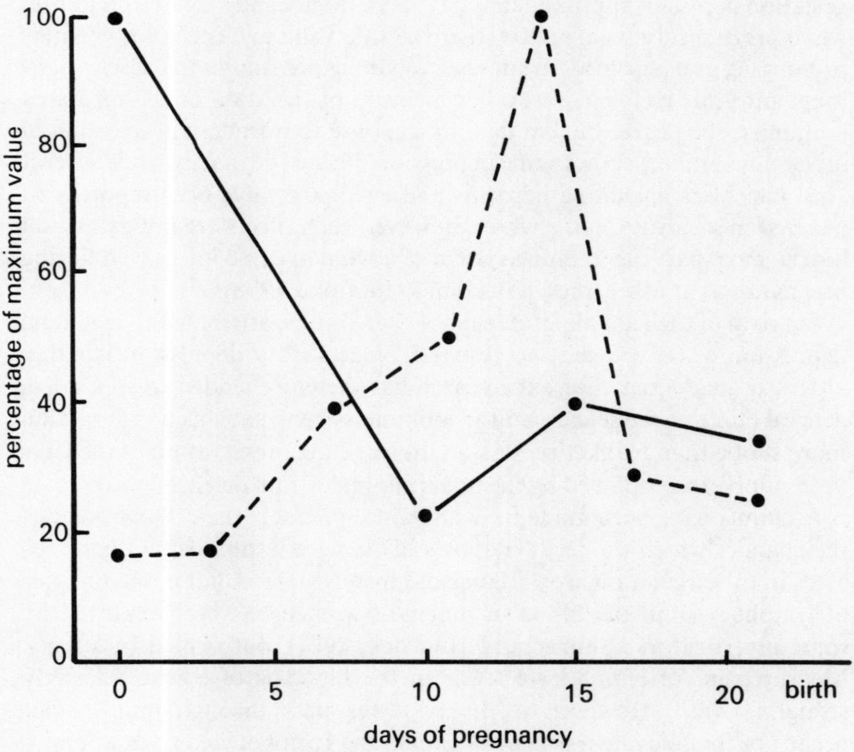

Figure 5.5 Variation in immune function during pregnancy in the mouse. Cell-mediated immunity (solid line) is assessed in terms of contact sensitivity to picryl chloride and humoral immunity (broken line) as response to sheep red blood cells as antigen; in each case the response is plotted as percentage of the maximum value obtained (redrawn from Fabris, 1973).

extend the survival of an allograft in a female which has previously been immunized against the graft donor (Beer and Billingham, 1976).

In contrast to rabbits, mice show no slowing of skin allograft rejection while pregnant, unless the skin donor is compatible with the pregnant female with respect to the major histocompatibility antigens (H-2) and differs from her only at minor histocompatibility loci. Even in these circumstances, the effect is slight. Allergic sensitivity to skin contact with picryl chloride—another form of cell-mediated immunity—has been shown to undergo some reduction in pregnant mice, but again the effect falls far short of a total failure to respond. Soluble antibody capacity, on the other hand, apparently rises during pregnancy (Figure 5.5)—see Fabris (1973); Fabris *et al.* (1977); Hamilton and Hellström (1977).

Deferment of rejection by a matter of a few days might conceivably be of some significance in a species with a short gestation period, but would be inconsequential in species with much longer pregnancies. Cattle, with a gestation period of approximately 280 days, do not show delayed rejection while pregnant. In women, too, there is little valid evidence of pregnancy prolonging graft survival; mean rejection times are similar to those in non-pregnant graft recipients. Another measure of the state of cell-mediated immunity, the degree of skin flare in response to intradermal injection of tuberculin antigen, shows some depression during pregnancy while several viral infections, including hepatitis and smallpox, have been reported to increase in severity. On the whole, however, such effects are not great, and for the most part the immune system of a woman can cope as well during pregnancy as at other times (Gusdon, 1976; Loke, 1978).

Survival of the fetal allograft cannot therefore be attributed to a general depression of the capacity to respond, since such a depression is either absent or inadequate. Recent research has therefore tended to seek a less general effect of pregnancy on the immune system, such as a modulation more subtle than blanket repression. In this case, preservation of the fetus need not be accompanied by the undermining of vital defences.

Attempts have been made to define more precisely the cellular basis of the changes which do occur. Reports of changes in numbers of lymphocytes in the circulation are variable and inconsistent. A fall in the number of lymphocytes in the blood of pregnant women has been reported by some investigators (Bulmer and Hancock, 1977), but denied by others. Measurement of lymphocyte levels in the blood is of course relatively straightforward. However, as lymphocytes pass through many other organs of the body, assessment of circulating lymphocytes gives at best a very narrow picture of any changes consequent on pregnancy. Other compartments of the lymphoid system may also alter in their lymphocyte content: the thymus undergoes involution during pregnancy, the spleen increases in weight and cell content, and circulating lymphocytes show a greater reluctance to leave regional lymph nodes. In short, there is a general redistribution of lymphocytes around the body rather than a simple fall in numbers (Beer and Billingham, 1976).

The composition of the circulating lymphocyte population also alters, although here too there are conflicting accounts of the nature of the alterations. In humans the number of T cells has been reported to decrease, and the number of B cells to increase, during pregnancy (Strelkauskas *et al.*, 1978). This might perhaps predispose a woman to develop blocking soluble antibody and specific immune enhancement, hence acceptance of her allogeneic fetus, rather than to mount a cell-mediated rejection response. The possible role of enhancement in pregnancy is discussed later (p. 124). However, several investigators have reported either a decline in B cell numbers, or no change in the relative numbers of B and T cells.

A decline in the proportion of circulating lymphocytes identified as T cells has also been reported in mice, during syngeneic as well as allogeneic pregnancies. Here, however, the proportion of B cells also fell. 'Null' cells, apparently carrying surface markers of neither T nor B cell type, rose sharply (Chatterjee–Hasrouni *et al.*, 1980).

The situation is therefore far from clear, with inadequacy of information about changes in lymphocyte distribution within the body compounded by some uncertainty about identification of lymphocyte subpopulations. There are no firm grounds for ascribing fetal survival to simple and consistent changes in lymphocyte numbers, or to the balance between T and B cells.

For a mother's immune response to her fetuses to be muted during pregnancy it is not necessary for lymphocyte numbers to alter. An alteration in sensitivity to stimuli, or in the capacity to respond, would suffice; as would a change in the balance between the different subsets of lymphocytes, such as a modulation of the relative numbers of helper and suppressor T cells (p. 15).

According to some reports, the response of human circulating lymphocytes to the mitogen phytohaemagglutinin (PHA) declines during pregnancy. In one case, for instance, lymphocytes from pregnant women were estimated to require a dose of PHA 3–4 times that needed to achieve the same degree of transformation with lymphocytes from women who were not pregnant. As PHA preferentially stimulates helper T cells, a reduction in the sensitivity of the cell-mediated immune system is probably indicated (Finn *et al.*, 1972; Fujisaki *et al.*, 1979; Stahn *et al.*, 1978; Walker *et al.*, 1972).

While many investigations lead to the same conclusion, some do not. Birkeland and Kristofferson, for example, exposed lymphocytes collected at different stages of pregnancy to a battery of mitogens and antigens, including concanavalin A (Con A), thought primarily to stimulate suppressor T cells, as well as PHA (Birkeland and Kristofferson, 1980*c*). In addition, B lymphocyte function was assessed by stimulation with a lipopolysaccharide extract of *Escherichia coli*, *E. coli* LPS, which is thought to act as a mitogen exclusively for B cells (Birkeland and Kristofferson, 1980*a*). A further attractive feature of their experimental design was that all samples of lymphocytes were frozen on collection, accumulated until after the series was complete, and then all processed at the same time, so eliminating fortuitous variations in the assay procedures and allowing '*in vitro* playback of the immunological course of pregnancy' (Birkeland and Kristofferson, 1980*b*).

Some of the results are shown in Figure 5.6. No significant change was found in the response to PHA at any time during or after pregnancy. The response to Con A fell sharply in early pregnancy, then rose again as delivery approached. These results suggest that suppression of the maternal response to fetal antigens by suppressor T cells is unlikely to be a significant factor in early pregnancy, although it may become so later.

The B cell response to *E. coli* LPS showed a slight rise at the second month of pregnancy, compared with values obtained with lymphocytes sampled before pregnancy, followed by a steady decline until delivery. Curiously, the T cell response to Con A and the B cell response to *E. coli*

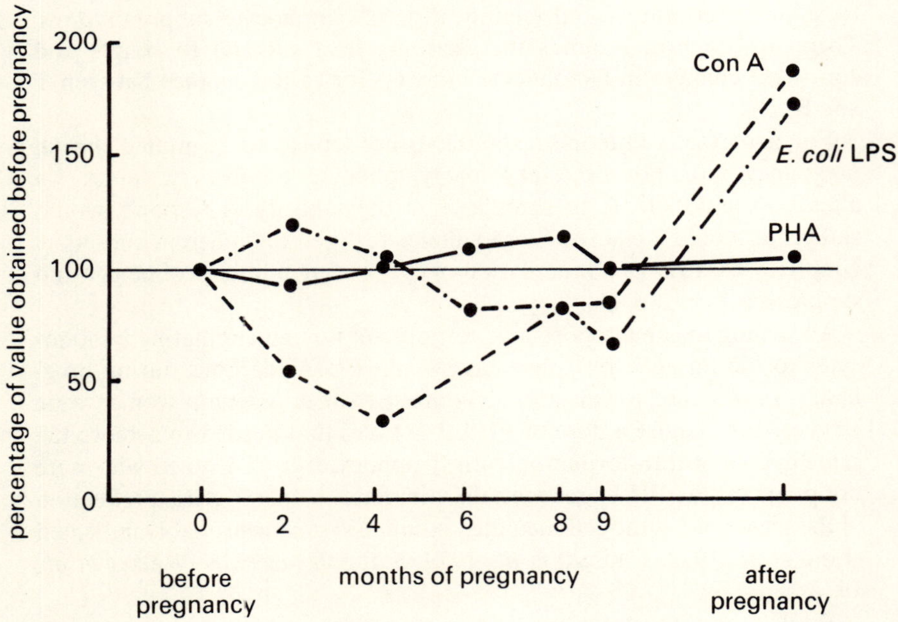

Figure 5.6 Response of human lymphocytes, at different stages of pregnancy, to the mitogens phytohaemagglutinin (PHA), concanavalin A (Con A), and *Escherichia coli* lipopolysaccharide (*E. coli* LPS). PHA stimulates T lymphocytes, Con A preferentially stimulates suppressor T cells, and *E. coli* LPS affects only B cells. All responses are plotted as percentage of the values obtained with lymphocytes collected before the start of pregnancy (redrawn from data in Birkeland and Kristoffersen, 1980*a*, 1980*c*).

LPS showed sharp rises with samples taken 3 to 5 months after delivery; in both cases the maximum values obtained were from these post-partum samples.

Variations in response of the peripheral blood lymphocyte population as a whole reflects two independent variables: alterations in the threshold of response of individual cells, and variation in the relative numbers of lymphocytes belonging to the relevant subsets of the lymphocyte population. Attempts have been made to separate these two variables by assessing the proportion of the different cell types in a way which does not depend on their responsiveness to stimuli.

In one study, B cells were identified by their surface receptors for the Fc region of IgG, and T cells by the convenient propensity to form spontaneous attachments to sheep erythrocytes. T cells were further subdivided according to whether they carried surface receptors for IgM or IgG; the former are thought to indicate helper T cells, and the latter suppressor cells (Hirahara *et al.*, 1980).

On this basis, no change was found during pregnancy in the relative numbers of T and B cells within the circulating lymphocyte population, although T cells declined (and B cells rose) in samples taken within four weeks after birth. Within the T cell population, suppressor T cells increased slightly during pregnancy, while helper cells fell sharply in the first third of pregnancy, and thereafter rose towards non-pregnant levels.

In mice, responses to both PHA and Con A have been found to decline in pregnancy. This was found to be true particularly of lymphocytes collected from the para-aortic lymph nodes; that is, those most directly concerned with the lymphatic drainage of the uterus, and immunological monitoring of its contents (p. 56). Cells from the inguinal lymph nodes showed a more variable depression of the response to mitogens, while cells from the more distant axillary and brachial lymph nodes showed no depression at all.

The generation of cytotoxic lymphocytes from their precursors is also locally suppressed within the para-aortic lymph nodes of pregnant mice, compared with the same lymph nodes of virgin mice, and the spleens of virgin and pregnant females (Clark and McDermott, 1978; Gottesman and Stutman, 1978). The response of lymphocytes to pregnancy appears therefore to depend not just on cell type, but also on position within the body; effects local to the uterus might prove decisive, and alterations in the behaviour of circulating lymphocytes comparatively irrelevant.

There is therefore good evidence that lymphocyte function is altered by pregnancy in humans and mice, although the nature of the changes has not yet been very clearly defined. Presumably they are caused by some stimulus emanating from the conceptus. What might this be?

The blood plasma of pregnant women inhibits the mixed lymphocyte reaction (MLR) (p. 24) when added to unrelated lymphocytes *in vitro*

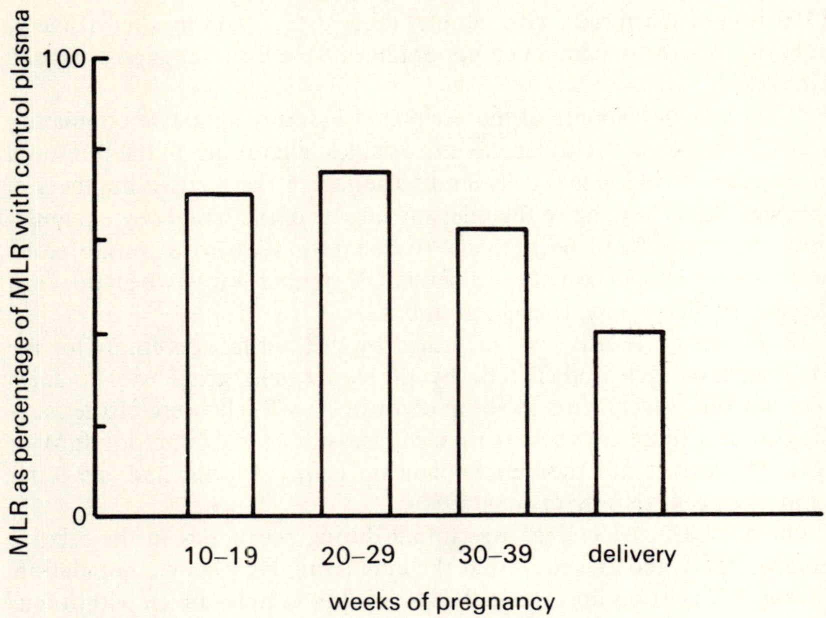

Figure 5.7 Inhibitory effect of blood plasma, collected at various stages of pregnancy, on Mixed Lymphocyte Reactions (MLR). Each is expressed as a percentage of the MLR obtained using control plasma from women who had never been pregnant (redrawn from Kasakura, 1973).

(Figure 5.7), and also the response to PHA (Bissenden *et al.*, 1980; Hill *et al.*, 1973; Schiff *et al.*, 1975). These findings implicate a soluble substance, or substances, present during pregnancy and with a suppressive effect on T lymphocyte activation. A number of hormones are specific to pregnancy, and it is among these that a non-specific immunodepressant has been most vigorously sought.

The most obvious candidates are steroid hormones: adrenal corticosteroid levels in the blood rise during pregnancy, while oestrogens and progesterone are produced in large quantities by the ovaries of pregnant females, and in many species by the placental trophoblast as well (p. 54). In humans, the placenta is in addition the source of two glycoprotein hormones: human chorionic gonadotrophin (HCG) and human chorionic somatomammotrophin (HCS). Analogous chorionic gonadotrophins have been identified in a number of other species. Of the two human placental glycoprotein hormones, HCG helps to maintain the corpus luteum during the first two months or so of pregnancy, but is present at a low level throughout pregnancy. HCS is involved with the control of fetal growth and maternal metabolism as well as preparation of the mammary glands for lactation.

All of these hormones have, in addition, immunosuppressive properties.

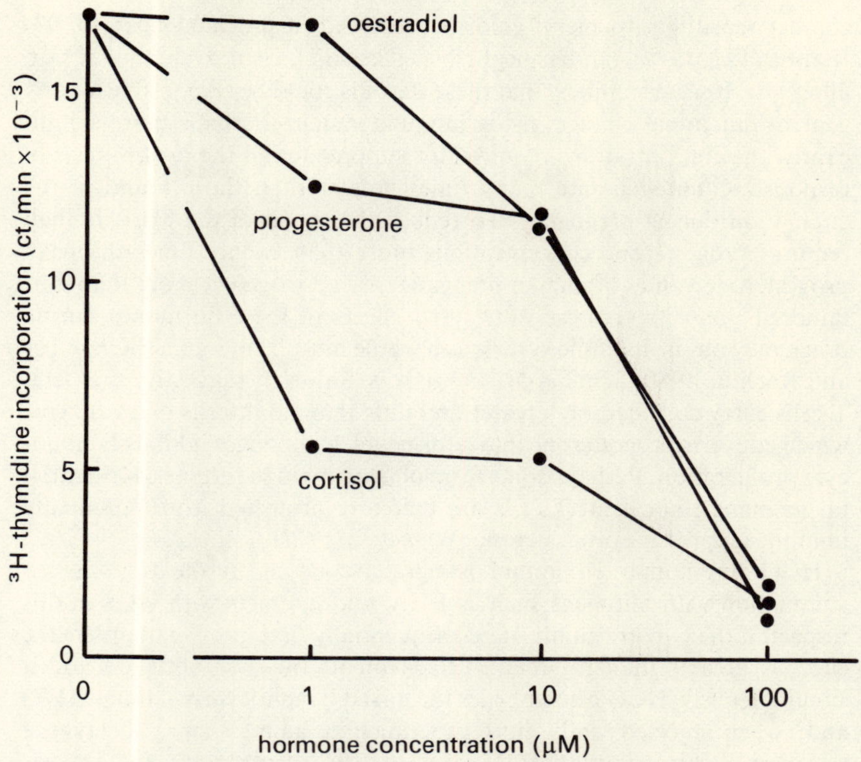

Figure 5.8 Effects of varying concentrations of oestrogen (oestradiol), progesterone and cortisol on the rate of uptake of ^{3}H-thymidine (a measure of DNA replication) by human lymphocytes following stimulation with the mitogen PHA (redrawn from Mendelsohn *et al.*, 1977).

Figure 5.8 shows the effects of progesterone, oestradiol-17β (the dominant oestrogen) and cortisol on the response to stimulation with PHA (Mendelsohn *et al.*, 1977). Similar depression by steroids of the response to Con A, and of the mixed lymphocyte reaction, can also be demonstrated (Clemens *et al.*, 1979; Pavia *et al.*, 1979; Wyle and Kent, 1977).

Although steroid hormones may be potent immunosuppressants in test systems *in vitro*, evidence of a significant immunosuppressive role *in vivo* is far from convincing. Injected corticosteroids mimic some of the effects of pregnancy on the lymphoid organs, such as involution of the mother's thymus, and may prolong the survival of skin allografts by a certain amount (Medawar and Sparrow, 1956). However, it has been shown that the ability of human blood plasma to inhibit the MLR does not correlate with its content of cortisol. This virtually excludes cortisol from being the major contributor to depression of immune function during pregnancy (Kasakura, 1973).

Oestrogen and progesterone treatment does mimic the depression of contact sensitivity to picryl chloride observed in pregnant mice (p. 93) (Fabris *et al.*, 1977), but has negligible effects on the survival times of skin allografts. It seems unlikely that these steroids could be responsible for any general functional change in the immune system. Perhaps more significantly, the concentrations required for suppression in the *in vitro* tests of response to mitogens are many times greater than those found in the circulation during pregnancy. To reduce the extent of the MLR by half requires progesterone concentrations more than twenty times the peak progesterone values of human pregnancy, or oestrogen at more than four hundred times its greatest value. Any effects of these hormones during pregnancy on the immune system as a whole must be marginal (Kitzmiller and Rocklin, 1980; Schiff *et al.*, 1975). It is, however, suggestive that fetal T cells carry considerably greater amounts than adult cells of an enzyme which converts progesterone into a form which no longer inhibits lymphocyte proliferation. Perhaps some form of fetal immune response is essential in pregnancy, and fetal T cells are therefore protected from potentially immunosuppressive progesterone (Weinstein, 1977).

HCG has a marked immunosuppressive effect on the response to stimulation with mitogens such as PHA, and interacts with HCS in this respect. If the two are administered sequentially (irrespective of order) the effect is greater than if either is given on its own, or both are added simultaneously. HCG also impedes the mixed lymphocyte reaction (MLR) and, when injected daily into experimental animals, may delay the rejection of skin allografts (Adcock *et al.*, 1973; Contractor and Davies, 1973; Hammarström *et al.*, 1979).

Two general modes of action have been proposed to explain the immunosuppressive actions of HCG. It might bind to the surface of allograft cells and mask their histocompatibility antigens so that recognition by patrolling lymphocytes does not occur; masking of antigenic sites by HCG is thought to play some part in the survival of some tumour cells, and of trophoblast (p. 145)—see Borland *et al.* (1975); Wiley (1980). Alternatively, it may bind to the surfaces of lymphocytes and so impede their interactions with other cells (including, in the case of the MLR, other lymphocytes).

This latter proposal is highly speculative. It is known that circulating lymphocytes during early human pregnancy tend to increase in electrophoretic mobility, due probably to an increased number of more highly mobile T cells. An increase in electrophoretic mobility indicates an increase in negative surface electrostatic charge, such as would result from the acquisition of a surface coating of a sialic acid-rich substance such as HCG. Incubation of lymphocytes *in vitro* with HCG causes such an increase in electrophoretic mobility. Finally, the period of enhanced electrophoretic mobility during pregnancy roughly corresponds with the period when HCG circulates at high levels (Loke *et al.*, 1980).

Increased negative surface charge of lymphocytes will make close mutual contact with cells of similar charge less likely and may therefore impede the interactions necessary for the MLR, or other types of immune response, to develop; by the same argument, recognition and destruction of trophoblast might be prevented. Two powerful objections can be raised to this proposed role of HCG in protecting the fetus. Firstly, the concentrations required to ensure prolongation of allograft survival are greater than those prevailing during pregnancy; and, secondly, the immunosuppressive effects of HCG diminish the purer the hormone preparation being used. The immunologically active component is therefore not the same as the hormone, but must presumably be a substance whose chemical and physical properties are sufficiently similar for crude extraction procedures to fail to separate the two. Much the same is true for HCS; purified HCS has relatively little effect on lymphocyte transformation induced by either PHA or allogeneic cells. The contaminating factors with immunosuppressive properties have not been identified, but appear to have molecular weights in the region of 20 000 to 40 000 (Gundert *et al.*, 1975; Kitzmiller and Rocklin, 1980).

An escape from the paradox of the immunosuppressive potency of hormones *in vitro* and their relative ineffectiveness *in vivo* may come from consideration of the possibility of purely local action near the site of production. The most critical point for the allogeneic conceptus is the placental interface, where trophoblast and maternal tissues confront each other directly. It is precisely here that placental oestrogen, progesterone, HCG and HCS (and their elusive associate) will be at their greatest concentrations. Maternal lymphocytes in the area—obviously those most likely to respond to the alloantigens of the conceptus—are therefore faced with much higher concentrations of potentially immunosuppressive hormones than their counterparts elsewhere in the body. The equivocal evidence for the immunosuppressive effect of hormones does not rule out the possibility that they may be potent immunosuppressants in the immediate vicinity of the placenta (Amoroso and Perry, 1975; Borland *et al.*, 1975).

A local immunosuppressant effect has been investigated by the implantation of small silastic tubes wrapped round with cotton threads. These were implanted subcutaneously into rats: after a week, a well-developed granuloma—a form of immune response, usually to a persisting stimulus, involving its encapsulation by macrophages—had developed round each implant. When the silastic tubes contained progesterone, which diffused slowly out into its surroundings, no granuloma formed. Corticosterone similarly prevented granuloma formation, but cortisol and oestrogen were both ineffective (Figure 5.9)—see Beer and Billingham (1979); Siiteri *et al.* (1977).

Granuloma formation does not normally play a conspicuous role in

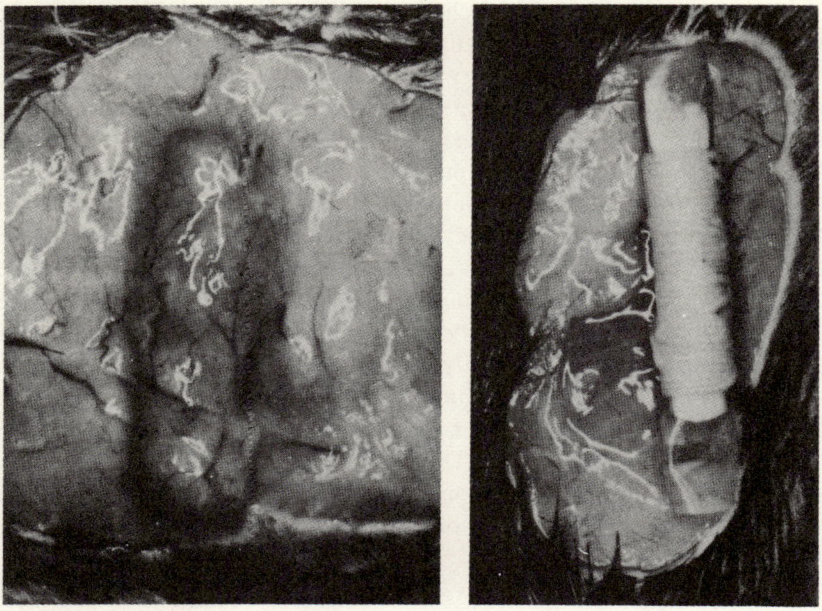

Figure 5.9 The local inhibitory effect of progesterone on granuloma formation. Both illustrations show cotton-wrapped silastic tubes 14 days after subcutaneous implantation into male rats; that on the right contained 100 mg progesterone (from Beer and Billingham, 1979, with permission).

allograft rejection, so a stimulus with a greater resemblance to a conventional allograft was produced by inserting silastic tubes into tubes of everted rat tail skin. In this case, progesterone delayed graft rejection and reduced the formation of circulating humoral antibody against graft alloantigens. Progesterone has therefore considerable potential as a local immunosuppressant.

It therefore appears possible that non-specific changes in the immune system take place as a result of pregnancy, that these are mediated, at least in part, by hormones, and that they may play at least an ancillary and local role in protecting the fetal allograft against rejection by its mother.

Other substances have also had immunosuppressive roles imputed to them. α-fetoprotein (AFP) is a glycoprotein found in mice, humans and other species, detectable (apart from some pathological circumstances) only during pregnancy. In human pregnancy it is produced mainly by the fetal liver, although the yolk sac contributes in the earlier stages, and rises to a peak of up to 4 mg ml^{-1} of fetal serum at 14 weeks of pregnancy. Thereafter, it declines steadily in concentration until birth. Passage across the placenta is meagre, so that concentrations in maternal serum are seldom more than 0.001 % of the maximum fetal levels (Lau and Linkins, 1976).

AFP is immunosuppressive when added to lymphocytes *in vitro*. Although its reported effects are very variable, it appears to act primarily on T cells and cell-mediated immunity. Mixed lymphocyte reactions are inhibited, as well as proliferative responses to PHA and Con A, and T-cell-dependent humoral antibody responses. AFP does not suppress the proliferative response to a mitogen specific to B cells (Murgita *et al.*, 1978; Murgita and Tomasi, 1975*a*, 1975*b*; Yachnin, 1975; Zimmerman *et al.*, 1977).

The mode of action of AFP is not known. Mouse AFP is polymorphic, with five components distinguishable. These differ in sialic acid content: only the fraction richest in sialic acid is immunosuppressive. Interestingly enough, a number of sialylated glycoproteins, including HCG, are also reported to have immunosuppressive properties (p. 100). AFP may bind to cell surfaces and 'blindfold' lymphocytes as HCG is thought to do; this would explain its potent suppressive action in lymphocyte interactions which involve response to determinants of the I region of the MHC (p. 19). It appears also to facilitate the development of suppressor T cells; a recent report suggests that HCG can do likewise, at least *in vitro* (Fuchs *et al.*, 1980). There is evidence that AFP's immunosuppressive property depends on its binding affinity for oestrogen; possibly it interacts in some way with steroid-induced immunosuppression (Murgita *et al.*, 1977; Peck *et al.*, 1978*a*, 1978*b*).

A number of serum proteins have been identified which are either unique to pregnancy or present at markedly raised levels during pregnancy, and which have immunosuppressive properties when tested *in vitro*. Two glycoproteins in particular appear significant. The first, an α-macroglobulin, is variously named pregnancy zone protein (PZP), pregnancy-associated globulin (PAG) and α_2-glycoprotein: these (and other) synonyms were given by different investigators who were probably independently working on the same substance. This has a molecular weight of 326 000. Although detectable in the serum of non-pregnant women (and in that of men), it is found at only low levels; during pregnancy it rises to a varying extent, and may exceed a serum concentration of 1 400 $\mu g\,ml^{-1}$ (Lin and Halbert, 1976; Stimson, 1980; von Schoultz *et al.*, 1973).

It is capable *in vitro*, at concentrations within the range of those occurring during pregnancy, of depressing T cell responses to mitogens and allogeneic cells. The similar α-globulin found in non-pregnant women has similar effects, and has been shown in addition to prolong allograft survival when injected into experimental animals. It may therefore have a general immunoregulatory function, brought into particular prominence by the needs of pregnancy.

The source of the substance is not clear. Peripheral blood leucocytes are capable of synthesizing it, but the cells involved appear not to be lymphocytes. Possibly monocytes are involved, since according to some reports these increase in numbers during pregnancy, and fetal monocytes have been shown to produce an unidentified substance with immunosuppressive properties (Wolf, 1977). Whatever the synthesizing cells, their activity appears to be stimulated by oestrogen.

A β_1-glycoprotein of molecular weight 90 000, designated SP1, appears to be specific to pregnancy. SP1 rises in the serum of pregnant women to a peak of 30 mg 100 ml^{-1} and falls sharply after delivery. It can inhibit the proliferation of PHA-induced lymphocytes, but not that of Con A-induced cells. Syncytiotrophoblast is probably the site of production, and immunohistology suggests an association with the plasma membranes of this tissue (although other studies have not confirmed this). Whether membrane-bound or not, SP1 like progesterone and other placental hormones may have a predominantly local action (Horne *et al.*, 1976).

It has been known for some time that the polyamines spermine and spermidine, in the presence of blood serum, have the ability to depress the response of lymphocytes to PHA. These substances are present in high concentrations in rapidly proliferating tissues, the placenta being particularly rich in spermine. (As also is seminal plasma, which has its own problems of immunosuppression.) When spermine is acted upon by a polyamine oxidase, substances are produced which are thought to inhibit cell division. The degree of suppression of lymphocyte transformation *in vitro*, when spermine is present, is proportional to the polyamine oxidase activity of the serum added. As serum polyamine oxidase levels have recently been shown to rise in pregnant women, there is a possibility that the action of polyamine oxidase on the locally high concentrations of spermine may contribute to immunosuppression within the placenta (Morgan and Illei, 1980).

Finally, an interferon (one of a group of proteins with potent intracellular antiviral properties) has been identified in the mouse placenta. Interferons have been shown to modulate a variety of lymphocyte responses to stimuli; injection of interferon into mice, for example, prolongs allograft survival (Fowler *et al.*, 1980).

There is, therefore, something of a plethora of candidates for the role of an immunosuppressant of pregnancy, among them steroid and glycoprotein hormones, pregnancy-specific glycoproteins, the polyamine/polyamine oxidase system, and others. The extent to which any or all of these are necessary for pregnancy to succeed has not been determined; indeed, the nature of the changes in the immune system which these are supposed to cause has not been fully defined (and will probably, in any case, differ from species to species). However, there is one feature most putative immunosuppressants have in common: if they emanate from the placenta, or are induced by some stimulus which does, they are unlikely to play a significant part until the placenta has become well established.

From fertilization until some time after implantation, the embryo would therefore get little or no help from any of the substances mentioned so far. (One exception to this may be HCG; a substance antigenically similar to HCG is present on the surface of preimplantation mouse blastocysts.)

It is therefore particularly convenient that a substance with immunosuppressive properties has been tentatively identified in the serum of mice, humans and sheep. It is designated *early pregnancy factor* (EPF), has a molecular weight of 200 000, and appears as early as 6 hours after conception. It persists in the circulation for the first six months or so of human pregnancy, although the initial EPF is replaced, from about 6 days after conception, by a variant of lower molecular weight. EPF inhibits various aspects of T cell function, and may therefore help to bridge the gap between fertilization, when paternal antigens on the embryo are first in danger of being recognized (see p. 107), and the establishment of a placenta sufficiently mature to reduce the effectiveness of the mother's immune system by other means (Clarke *et al*., 1980; Morton *et al*., 1976; Noonan *et al*., 1979). Despite the convenience of EPF, it has to be pointed out that not all investigators can find evidence of its existence, at any rate in humans (Cooper and Aitken, 1981).

The large number of immunosuppressive agents associated with pregnancy does present a somewhat embarrassing problem. With all these immunosuppressants in the circulation, how could a pregnant female ever mount an effective immune response to infection, or to an experimental allograft? And yet, pregnant females *can* reject allografts, albeit possibly after some slight delay; and, with few and minor exceptions, most infections are successfully combated during pregnancy.

If there is no generalized immunosuppression of pregnancy, the quest for a generalized immunosuppressant becomes pointless. It may be, therefore, that the substances discussed—whose immunosuppressive properties, in any case, have often been inferred from experiments *in vitro*, frequently with no corroborative evidence *in vivo* of immunosuppression—are largely irrelevant to the question of immunological survival of the fetus; or, if their immunosuppressive properties do have any relevance *in vivo*, that their effects are primarily local.

5.5 Is the fetus antigenic?

No question of immune rejection of the fetus would arise if, as Little perceptively suggested as long ago as 1924, it 'has no definite physiologic characteristics which are individual enough to be recognized as foreign to the mother'; that is, if expression of histocompatibility antigens of paternal derivation is deferred until their presence no longer mattered.

Injection of embryonic cells can accelerate rejection of a subsequent allograft of skin from an individual of the same genetic strain as the embryo. This indicates that embryonic tissues carry histocompatibility antigens, but does not conclusively establish their presence at the time of injection. Antigen expression, subsequent to the priming injection of cells but prior to receipt of the skin graft, might possibly be responsible for sensitization.

When trophoblast and inner cell mass cells are surgically separated from each other and independently transplanted to, for example, the kidney capsules of mice of a different genetic strain, trophoblast survives while the embryonic tissues do not. Apparently antigens sufficient to cause rejection are present on embryonic tissues but absent from trophoblast; in an intact embryo, trophoblast might therefore represent an antigenically neutral *cordon sanitaire* protecting the embryo from its mother (Edwards *et al.*, 1975; Kirkwood and Billington, 1981).

Again, the expression of antigens only after transplantation cannot be ruled out because of the time which elapses, after the transfer, before the recipients become sensitized and immune rejection advances far enough to be recognized histologically. If the recipients are intensively preimmunized against donor strain antigens, transplanted embryos (ranging from early cleavage stage to $3\frac{1}{2}$ day blastocysts) completely disappear within a week. This rapid destruction scarcely leaves time for antigen expression, and makes such expression only after transplantation seem a less plausible explanation (Searle *et al.*, 1974).

In view of the apparent antigenic nullity of trophoblast inferred from its survival when separately transferred, the destruction of whole blastocysts in preimmunized hosts is surprising. This need not imply an immune reaction against trophoblast itself; for trophoblast outgrowth and successful implantation, an inductive stimulus from the inner cell mass is probably essential, and the failure to develop might therefore result from antibody-mediated damage to the inner cell mass. Antibody can be detected within the interior of mouse blastocysts; as it cannot at this stage of development be endogenous, early embryos are clearly permeable to maternal immunoglobulin molecules (Bernard *et al.*, 1977).

The antigens responsible for provoking the recipient into rejection have been identified by using strains of mice differing at known genetic loci, again using intensively preimmunized recipients. Some results are summarized in Table 5.3. When $3\frac{1}{2}$ day blastocysts were transferred from mice of a genetic strain differing from that of the recipient only with respect to minor histocompatibility antigens and compatible at the major H-2 locus, little development took place after transfer. Infiltration of the embryos by polymorphonuclear leucocytes and lymphocytes started within two days, and by the 6th day destruction was total. When the zonae pellucidae (p. 51) of the embryos were enzymically removed before transfer, invasion of host leucocytes was apparent within 24 hours. Such a rapid attack indicates that the eliciting antigens were most probably

Table 5.3 Percentage survival of $3\frac{1}{2}$ day mouse blastocysts after ectopic transfer to presensitized and naïve recipients. Donor blastocysts were either of the same strain as the recipients or differed from them with respect to both H-2 and non-H-2 antigens, H-2 alone, or non-H-2 alone. When the donor and recipient were syngeneic, recipients had previously been sensitized to cells of a 'third-party' strain; with H-2 disparity alone, immunization was to cells of a third strain identical to the blastocyst donor strain at the H-2 locus (data from Searle *et al.*, 1974).

	State of recipient	
Disparity	*Naïve*	*Presensitized*
None (syngeneic)	—	89%
H-2+non-H-2	78%	9%
Non-H-2	100%	10%
H-2	—	87%

detectable at the time of transfer, and not expressed subsequently (Searle *et al.*, 1974).

Disparity only of minor histocompatibility loci is therefore sufficient for blastocyst rejection, and these antigens appear to be expressed on $3\frac{1}{2}$ day blastocysts.

When donor and recipient were identical at the minor histocompatibility loci but differed at the major, H-2, locus the transplanted blastocysts survived and developed normally. An H-2 disparity is, of course, quite adequate to ensure rapid rejection of a graft of adult tissue. The major histocompatibility locus does not therefore appear to be effectively expressed on mouse blastocysts.

This conclusion is supported by experiments in which mouse blastocysts, deprived of their zonae pellucidae, were treated *in vitro* with complement and antisera prepared in mice differing either at an assortment of histocompatibility loci, or only at the H-2 locus. Two days later, embryos in the former case were dead; in the latter treatment they had survived unscathed (Heyner *et al.*, 1969).

Both sets of experiments therefore indicate that the minor histocompatibility antigens—those not normally responsible for rapid graft rejection—are present on early mouse embryos, while the more significant antigens of the H-2 locus are not in evidence.

Antisera specifically directed against particular histocompatibility antigens, in conjunction with a range of labelling methods, have been used in attempts to clarify the situation further. These methods confirm that antigens other than H-2, including H-3, H-6 and, in male embryos, H-Y, are expressed from an early stage of development. By the 4 or 8 cell stage, antigens of paternal derivation are definitely included. Apart from a decline in antigenicity at around the time of implantation, all tissues of the conceptus thenceforth carry significant amounts of non-H-2 antigens (Billington *et al.*, 1977; Edwards *et al.*, 1975; Heyner *et al.*, 1980; Sellers *et al.*, 1978).

Most serological evidence indicates that H-2 antigens are absent, or present in only negligible amounts, on cleavage stage embryos and morulae in normal pregnancies. Shortly before implantation, low levels of H-2 antigen are detectable but thereafter the trophoblast reverts to its non-antigenic status with respect to H-2; non-H-2 antigens may still be expressed. Cells derived from the inner cell mass, in contrast, develop and continue to express major histocompatibility antigens as well as those of the minor H-loci (Billington *et al.*, 1977; Carter, 1978; Chatterjee-Hasrouni and Lala, 1979; Edwards *et al.*, 1975; Jenkinson and Owen, 1980; Muggleton-Harris and Johnson, 1976; Palm *et al.*, 1971; Searle *et al.*, 1976; Sellens *et al.*, 1975).

If both ovaries are removed from a female mouse after fertilization but before implantation, the blastocysts in the uterus enter a resting state. This may persist for many days, providing only that a daily injection of progesterone is supplied to support the uterus, in the absence of corpora lutea. Resumption of embryonic development, and implantation, rapidly follow a single injection of oestrogen, a procedure which mimics an action which would have been carried out by the ovaries in an intact female.

Blastocysts in the resting state are markedly antigenic compared with those of an undelayed pregnancy, and carry significant amounts of H-2 antigen of paternal origin. When oestrogen causes development to resume and implantation to start, this antigenicity falls sharply; 14 hours later, H-2 antigens are no longer detectable (Hakansson and Sundqvist, 1975; Håkansson *et al.*, 1975).

The survival during delay of blastocysts which express antigens suggests that until implantation there is little risk of rejection; the dangers presumably increase with implantation, and a decline in antigenicity at this time may be an important element in the negotiations between uterus and blastocyst. Delayed blastocysts reveal in a more obvious form a situation which obtains also in undelayed pregnancies.

Delayed implantation does not result only from experimental manipulations. Mice frequently conceive immediately after giving birth. The mother would therefore have to bear the metabolic load of supporting two broods simultaneously, one in the uterus and the other suckling. The load is lessened by postponing implantation of the newly conceived litter—so deferring the metabolically costly phase of pregnancy—until the demands of the earlier litter have somewhat abated. Resting blastocysts during this lactational delay of implantation are in a very similar state to those in artificially induced delay, so the experimental findings probably accurately reflect a situation which arises frequently in normal circumstances (Austin and Short, 1972*b*).

During the early preimplantation period, before H-2 antigens are expressed, further antigens are present, some peculiar to embryos and not present in adult cells. One of these, known as F9, is a member of a class of 'carcinoembryonic antigens', found only at specific stages of embryonic cells and in certain forms of tumour in adults. The expression of F9 and of H-2 antigens appear to be mutually exclusive, although nothing is known of how such a reciprocal control operates; as H-2 appears, F9 disappears from all embryonic cells except those which in male embryos are destined to give rise to germ cells (Gachelin *et al.*, 1976).

F9 has been tentatively identified as a product of the T locus, which is known to be involved in some way in cellular recognition processes during embryogenesis. Anti-F9 antibody *in vitro* inhibits an episode of cell aggregation which is essential to the normal development of early mouse embryos, while immunizing female mice against the antigen drastically reduces fertility (Ducibella, 1980; Hamilton *et al.*, 1978).

Although little comparable work has been carried out on other species, the situation with respect to the major histocompatibility antigens in rats appears broadly similar to that in mice. Antigens of the MHC (in this case termed Ag B) are deficient in trophoblast, and only minor histocompatibility antigens are present (Ferguson and Palm, 1977; Sundqvist *et al.*, 1977).

Fetal cells are therefore antigenic, and the surrounding trophoblast apparently at most only weakly so. Before concluding that the fetus survives, partly at least, by hiding behind its trophoblast, the *degree* of antigenicity of fetal cells must be assessed. Although detectable, antigens might be present in amounts so small that recognition and destruction might be impaired. (In the light of experiments already discussed, however, any such contribution to fetal survival could only be slight.)

If the immune system of adult mice is weakened by X-irradiation, skin allografts from newborn mice are accepted for longer periods of time than genetically identical adult skin. Given sufficient radiation, and disparities only at minor histocompatibility loci, permanent acceptance may even occur. Even with H-2-incompatible grafts, skin from newborn mice survives roughly twice as long as adult skin. Similar results have been obtained with hamsters and humans, using in the latter case byproducts of circumcision operations. By extension, fetal skin may likewise be less antigenic than the skin of adults, and the risk to the fetus might thereby be reduced; the benefit is, however, likely to be slender (Billingham and Silvers, 1964; Wachtel and Silvers, 1971).

In mice, trophoblast is not the only tissue of the conceptus which abuts maternal tissues. In later pregnancy, the outer face of the endoderm of the conspicuous yolk sac is also exposed in the uterus (Figure 5.3). When yolk sac epithelium is transplanted as an allograft, it is rapidly rejected, so can be presumed to carry major histocompatibility antigens. Careful histology combined with the use of specific antisera suggests, however, that histocompatibility antigens are not uniformly distributed over all surfaces of the yolk sac cells, but are restricted to the fetal side of the yolk sac membrane. The side of the membrane towards the uterine lumen, separated by a zone of occluding junctions, is apparently devoid of histocompatibility antigens. The face the yolk sac presents to the uterus is therefore antigenically neutral. Possibly cells of other tissues exposed to the mother, including the outer layers of trophoblast, might similarly show polarized expression of antigens, and so contrive to express surface antigens in the normal way while maintaining an acceptably neutral outer face (Parr *et al.*, 1980).

The question of the antigenicity of trophoblast, particularly of trophoblast of the mature established placenta, is sufficiently complex to require further discussion (see section 6.6). For the present, it is sufficient to conclude that, although the regions of direct contact between the conceptus and the mother—trophoblast and, in the mouse, the yolk sac membrane—are of low antigenicity, the embryonic tissues within express significant amounts of both major and minor histocompatibility antigen. It may be that the trophoblast is a barrier, impenetrable to fetal cells and maternal lymphocytes, which protects the embryo within.

5.6 Is there a barrier between mother and fetus?

The fetus itself is antigenic, and the mother's capacity to mount an immune response is not drastically diminished by pregnancy. It is thus reasonable to postulate that rejection of the fetus by its mother might be avoided by some form of barrier interposed between the two. This could interdict either afferent passage of antigen—in soluble or cellular form—to the mother, or the efferent passage of soluble antibody or sensitized effector cells into the fetus.

For much of the period between fertilization and implantation, the mammalian embryo is surrounded by a noncellular zona pellucida. This is not totally non-antigenic; autoantibodies to zona antigens have been reported occasionally in women, and may contribute to infertility (p. 81). It is, however, probably only weakly antigenic and must for the most part be insufficiently so to cause any immunological problems.

The zona pellucida might well prevent egress of antigenic material from the embryo. However, this is unlikely to be enough to prevent an immune response, since even in her first conception a female will already have been exposed to spermatozoa carrying paternal alloantigens. In mice and humans, seminal exposure has been shown to result in some development of cell-mediated immunity (McLean *et al.*, 1980). There is little evidence of any harm coming to embryos as a result, although in view of the scale of the mortality of preimplantation embryos, this cannot altogether be ruled out. As an afferent barrier, the zona pellucida is not likely to be a significant factor in survival of the conceptus.

Despite some earlier reports to the contrary, there is now good evidence that immunoglobulins (including molecules as large as IgM), and all necessary components of complement can freely penetrate mouse blastocysts with their zonae in place. In fact very little immunoglobulin is actually found within preimplantation embryos, although the zona is no impediment to entry, possibly because much of the immunoglobulin which does enter is rapidly digested by the embryo (Bernard *et al.*, 1977).

As already mentioned (p. 106), the zona does delay leucocyte infiltration of blastocysts transferred into previously hyperimmunized hosts. This is unlikely to be relevant to normal development, since in most species the zona is shed before implantation and an embryo is unlikely to be threatened by a cell-mediated attack while still free in the oviduct or uterus. Kirby found that zona-free mouse blastocysts could survive for at least 14 days after transfer into the uteri of previously immunized recipients, and were then still capable of implanting and developing normally. By the time they have developed into blastocysts, embryos are surrounded by trophoblast and have no need of a zona for protection. If the zona pellucida has any significance as an immunological barrier, whether in preventing the egress of antigen, or the entry of leucocytes, it can be no more than minor and transient (Heyner *et al.*, 1969; Kirby, 1968, 1970; Searle *et al.*, 1974).

Trophoblast appears to be deficient in antigens, at least those of the major histocompatibility locus (p. 107; but see also p. 135). Could it constitute a barrier between an antigenic fetus and its immunologically competent mother?

If fetal transplantation antigen does pass across the putative barrier of placental trophoblast in soluble form, the amounts involved are too small to be directly detectable by currently available methods. The afferent passage of fetal cells, on the other hand, can be established rather more readily. In humans, fetal erythrocytes (red blood cells) can be distinguished from adult erythrocytes by a variety of features; among these is a specifically fetal form of haemoglobin (Hb F) which can readily be distinguished from the normal adult haemoglobin, Hb A. On this basis, up to about 50% of women carry fetal erythrocytes in their circulation after giving birth (Figure 5.10). In many cases the transfer of fetal erythrocytes takes place in the final stages of pregnancy, no doubt facilitated by the disruption of placental blood vessels as birth approaches and by the blood pressure gradient between mother and fetus. Even in early pregnancy, however, some transplacental passage takes place; in one survey, fetal erythrocytes were found in 6.7% and 15.9% of women during the first and second trimester of pregnancy, respectively (Beer and Billingham, 1976; Loke, 1978; Scott and Jones, 1976).

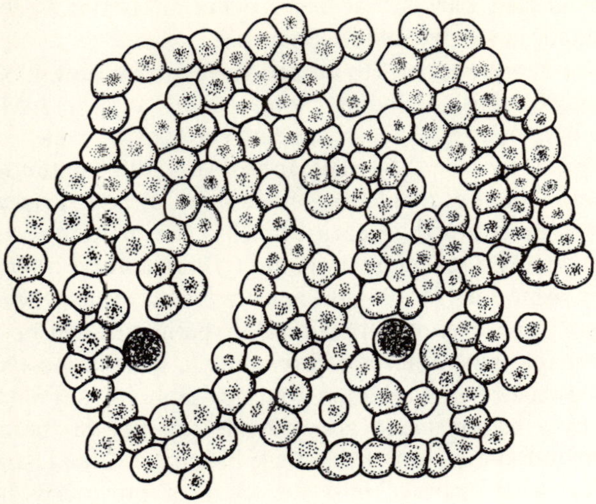

Figure 5.10 Blood smear taken from a woman during the ninth month of pregnancy, and treated with an acid elution technique which reveals two erythrocytes containing fetal haemoglobin (redrawn from Beer and Billingham, 1976).

Human erythrocytes may to some extent be a special case, since they lack major histocompatibility antigens. Sensitization against blood group antigens does occur (p. 146), but is rare considering the high frequency of blood group incompatible pregnancies.

Leucocytes, on the other hand, do carry MHC antigens and are fully capable of eliciting an immune response. Fetal leucocytes are less easy to identify in the maternal circulation than fetal erythrocytes. One approach is to look for cells with a Y chromosome among the cells circulating in mothers carrying male fetuses; the Y chromosome can fairly readily be identified by its intense fluorescence when stained with quinacrine dye. This made it possible to demonstrate fetal cells in 42 out of a sample of 45 women pregnant with what were subsequently discovered to be male fetuses. A few women carrying female fetuses (of sex chromosome constitution XX) do have leucocytes apparently with fluorescing Y chromosomes; these might have persisted from previous pregnancies, or might be due to cytological misinterpretation. Uncertainty about the method is also suggested by some conflicting reports of failure to find any Y-containing cells in pregnant women (Loke, 1978).

In mice, the long-term colonization of maternal lymphoid organs by fetal cells has been reported. In some cases, fetal cells have been found in maternal spleens as long as two months after delivery of the fetuses from which they came. Although the fetal cells were capable of proliferating, and appeared to be functional lymphoid cells, there appear to have been no ill effects on the mother, on subsequent litters, or on the cells themselves (Gaillard and Liegeois, 1980). Fetus-versus-mother responses may occasionally occur, but cannot often be of great significance (Chardonnens and Jeannet, 1980).

The evidence therefore suggests that fetal leucocytes of various sorts (as well as deported trophoblast fragments; see p. 86) enter the circulation of the mother during pregnancy, and that the placental barrier is not effective in an afferent direction.

Further supporting this view is the considerable mass of evidence of the immunological recognition by pregnant females of their genetically alien fetuses. In mice, guinea pigs, and other species, the regional lymph nodes which drain the uterus enlarge in response to pregnancy, and enlarge more with an allogeneic pregnancy than with a syngeneic one (p. 90); while both cell-mediated and humoral responses to fetal antigens have been well characterized in a number of species (p. 116). The picture that emerges is that the maternal immune response to the fetus is not a rare and pathological event normally avoided, but a routine (perhaps even necessary) consequence of pregnancy. The nature of this response is discussed in the next chapter. For the present, it is sufficient to conclude that, if a maternal response is normal, fetal survival cannot depend on the placenta being an afferent barrier to fetal antigens, although placental *limitation* of the afferent movement of fetal material is probably important.

Despite the lack of an effective afferent barrier, an efferent barrier could still prevent the entry of antibody or lymphocytes into the fetus. In fact antibody molecules enter the fetus relatively freely. The umbilical cord blood of the human fetus contains only small amounts of IgA and IgM, but strikingly high levels of all four subclasses of IgG (although IgG2 appears to be somewhat discriminated against). The concentrations found correlate closely with those in maternal circulation; this, together with other evidence, indicates that the antibody is predominantly maternal in origin, and that transmission into the human fetus starts by the 10th week of gestation. Other species studied, particularly the rabbit, also show fairly free access of antibody molecules into the fetus (Brambell, 1970; Chandra, 1976).

Specifically antifetal antibody, in contrast, is not detectable in the fetal circulation. Presumably its entry would rapidly be followed by adsorption to antigenic sites on fetal cells, so this is not very surprising. There is, in fact, reason to believe that most antifetal antibody is filtered out at the level of the placenta, and is unlikely to enter the circulation of the fetus itself. In some circumstances, however, entry of antibody against fetal antigen can be inferred. In rhesus haemolytic anaemia of the newborn, antibody against a fetal erythrocyte antigen causes severe anaemia (p. 146); the fact that this happens very largely towards the end of pregnancy emphasizes the relative impermeability of the barrier at other times. The antifertility effect of immunization against the F9 carcinoembryonic antigen (p. 109) suggests that here, too, antifetal antibody enters the embryo. The rarity of such effects suggests that antifetal antibody is normally excluded.

Selective filtering out of antifetal antibody probably takes place in the placenta, although as much of the passive transfer of antibody in the rabbit (for example) takes place across the yolk sac, other mechanisms may also be important. In fact any antifetal antibody which did enter would soon be absorbed from the circulation. Nor need it necessarily be cytotoxic. Human fetal lymphocytes, immediately after removal from the fetal circulation, are unaffected by cytotoxic antibody against HLA antigens of paternal type. With overnight incubation, paternal HLA antigens reappear on the fetal cells, and relevant anti-HLA antisera are now cytotoxic (Tiilikainen *et al.*, 1974). Perhaps the most likely explanation is that the antigenic sites of paternal derivation are masked by the attachment of antipaternal antibody which fails to fix complement, and is therefore not cytotoxic. The masking and protective antibody is shed overnight, re-exposing the antigenic sites. An alternative explanation, for which there is some evidence, is that the mother—possibly again by way of specific antibody—can modulate the phenotypic expression of MHC antigens by fetal cells, and reduce the expression of antigen 'foreign' to herself. The evidence for this being important in fetal survival is slight, but it cannot be totally dismissed as a possible contributor (Jacobs and Uphoff, 1974).

Even a slight reduction in density of alloantigenic sites on fetal cells might usefully reduce the cytotoxic effects of complement-fixing antibody, since complement assembly requires a certain minimum density of attached antibody molecules (Roitt, 1980). Deficiency of complement components in the fetal circulation would also reduce the likelihood of damage; the fetal complement system does appear to be in some respects relatively immature (Loke, 1978), although this probably has only a marginal effect.

Do maternal cells enter the fetus? Maternal erythrocytes are not unusual within the fetal circulation, but most evidence suggests the entry of maternal lymphoid cells to be rare. As XY cells in the mother indicate transplacental passage of cells into the mother, so XX cells in the blood of a male (XY) fetus show passage in the opposite direction. In only 1–2% of cases do human fetuses show the presence of maternal cells by this criterion (Adinolfi, 1975; Loke, 1978).

When maternal leucocytes were withdrawn, labelled with a fluorescent dye, and returned to the mother just before delivery, approximately two-thirds of the babies showed fluorescing cells in their umbilical cord blood on delivery. Assuming the dye does not separate from the originally labelled cells, traverse the placenta in solution, and reattach to fetal cells, this suggests that cell entry into the fetus may not be as rare as chromosome analysis suggests; possibly cells do enter the fetus quite frequently, but are rapidly destroyed.

Successful long-term colonization of the fetus by maternal cells is, however, possible. Female rats sensitized to skin allografts may in some cases transmit that sensitization to their offspring. Presumably this takes place by the movement of committed T cells across the placenta, although other mechanisms, for example the transplacental passage of specific soluble factors—see p. 20—are possible (Beer and Billingham, 1973).

Another example of the transplacental passage of maternal leucocytes, apparently a major exception to the normally restricted passage of cells into the fetus, has been reported in mice. When blastocysts of one strain of mouse were transferred into the uteri of foster females of another strain, large numbers of leucocytes of the foster strain (distinguishable by a chromosome marker) were later found to have entered the fostered fetuses. In some, as many as 34% of the cells of the fetal bone marrow were maternal in origin (Tuffrey *et al.*, 1969). Fortunately for the peace of mind of geneticists, the phenomenon does not appear to be a general one, as massive transplacental movements were not confirmed in later attempts to repeat the experiments (Billington *et al.*, 1969). Although the explanation in this particular case is not clear, it may hinge on some idiosyncrasy of the particular strain combination used; not all placenta are as leaky. Limited movements of cells do, however, occur; both between mother and fetus, and between sibling fetuses (Collins *et al.*, 1980; Gaillard and Liegeois, 1980).

Some natural examples of the successful long-term colonization of fetuses by maternal cells have been recorded. One human infant—a male, and therefore of sex chromosome constitution XY—was found to contain presumably maternal XX cells at the age of 39 weeks. In another case, infiltrated maternal cells may have been responsible for the death of another human infant; the symptoms were similar to those produced by injecting infant rats with lymphocytes sensitized against their own antigens, but against which they cannot themselves react (Kadowaki *et al.*, 1965). Cases such as these—with or without the complication of a graft-versus-host reaction by the foreign lymphocytes—are rare in the extreme. In general, the entry into the fetus of maternal cells other than erythrocytes is unusual, the number of cells involved probably small, and their survival brief. This may contribute to survival of the fetus despite maternal immune recognition of its presence.

CHAPTER SIX

THE NATURE OF THE MOTHER'S RESPONSE

THE FETUS IS ANTIGENIC, THE MOTHER IS IMMUNOLOGICALLY RESPONSIVE while pregnant, and responses to her fetus are not blocked by a simple placental barrier keeping asunder what conception has brought together. Fetal survival cannot therefore be explained by a failure of the mother to recognize her pregnancy. A variety of evidence (some of it already mentioned) indicates that immune recognition of the fetus by its mother does routinely occur. Why then is the maternal immune response not a destructive one? To attempt to answer this question, it is necessary to consider in more detail the nature of the maternal reaction to antigens of the conceptus.

6.1 Humoral antibody responses to the fetus

In pregnant women, serum antibody against fetal antigens of the major histocompatibility system (HLA) is common, to the extent of providing a useful source of reference sera for the tissue-typing of potential organ transplant donors and recipients. Some 12% of women, while pregnant, form detectable amounts of cytotoxic serum antibody which reacts with antigenic determinants carried by leucocytes of the father, and about 25% form corresponding agglutinating antibody. As pregnancy advances, the likelihood of such antibodies being formed generally rises. Increasing parity also brings an increase in the incidence of antipaternal antibody in the circulation. Estimates of the incidence depend on the precise criteria employed, but as many as 60% of women have been reported to have such antibodies in the circulation during their fourth or subsequent pregnancies (Doughty and Gelsthorpe, 1974, 1976; Loke, 1978; Salinas *et al.*, 1975).

In addition to antibody against the classical transplantation antigens of the MHC, there is evidence of antibody formation against paternal alloantigens which appear to be homologous with the Ia antigens of mice (products of the immunoregulatory Ir region of the MHC). This may be of particular relevance to the question of fetal survival, since Ia antigens appear to be excellent inducers of tolerance both to themselves and to certain accompanying conventional antigens (section 2.4)—see Ferrone *et al.*, 1976; Streilein and Klein, 1980; Winchester *et al.*, 1975.

There have also been reports of the natural occurrence during human pregnancy of antibody against isoantigens unique to embryonic tissues (or shared by embryonic and some tumorous tissues).

Similar circulating antibody against relevant paternal alloantigens has been found in other species, including cows and mice. In certain inbred strains of mouse, antibodies against MHC antigens (in this case H-2) are detectable as haemagglutinins; this is because mouse erythrocytes, unlike human ones, carry MHC antigenic determinants. The incidence of antifetal haemagglutinins varies unpredictably both between individual pregnant females and within an individual. Antibody may appear in a first pregnancy, be absent during a second, and reappear in a third; or disappear in the course of pregnancy and reappear after birth. Fertility is unaffected by such fluctuations (Goodlin and Herzenberg, 1964; Newman and Hines, 1980).

It has recently been shown that the preponderant antibodies induced by pregnancy to paternal alloantigens belong to the IgGl class. In mice these are not complement-fixing (or cytotoxic) and might therefore impede cytotoxic antibody by competitive exclusion, or act as enhancing antibody in relation to a cell-mediated response; see section 6.2 for a further discussion of this possibility (Beer and Billingham, 1980).

Circulating antibody may also be induced against minor histocompatibility antigens of the fetus. In both mice and rats, antibody against H-Y antigen has been found in pregnant females; in the latter species, with an incidence of 52%. Anti-H-Y appears to be able to cross the placenta, since it was detectable in 45.5% of the female offspring. Its apparent absence in male offspring was presumably due to absorption, as soon as it entered fetal tissues, by cell-bound H-Y antigenic determinants which females lack. Neither fertility nor sex ratio were affected (Krupen-Brown and Wachtel, 1979; Shalev, 1980).

Measurement of serum antibody levels is a fairly crude assessment of the immune response to a fetus. Many factors can modulate antibody secretion and disappearance rates. A more direct indication of B-cell response is the extent of rosette formation by maternal lymphoid cells when mixed in culture with paternal strain erythrocytes. A rosette consists of a lymphoid cell surrounded by a number of adherent erythrocytes, and normally forms only when the lymphoid cell carries on its surface immunoglobulin specific to antigenic determinants of the erythrocyte cell membrane.

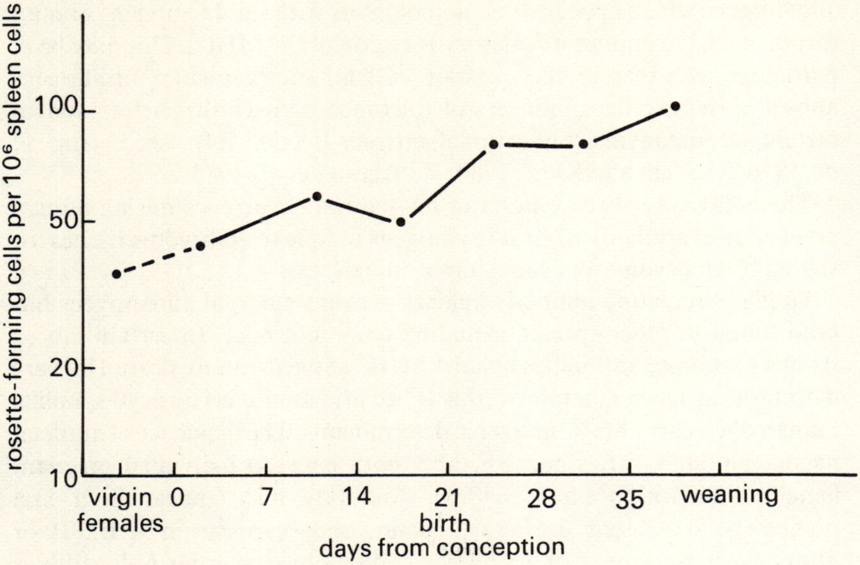

Figure 6.1 Sensitization of female mice against paternal antigens. Maternal antibody-forming cells directed against paternal erythrocytes, when mixed *in vitro* with erythrocytes of the paternal strain under appropriate conditions form rosette-like aggregates in which a maternal cell is surrounded by a ring of attached paternal cells. The proportion of maternal lymphoid cells which form such rosettes indicates the degree of sensitization of the maternal humoral antibody system to paternal antigen (based on data from Baines *et al.*, 1976).

As Figure 6.1 shows, during a (fifth) H-2 incompatible pregnancy, the proportion of rosette-forming cells in the spleen cell population rises progressively. In pregnancies which follow mating between individuals compatible with respect to the MHC but incompatible, to differing extents, at other histocompatibility loci, the results are more variable; in some cases rosette-forming cells increase during pregnancy, in others they do not. Only a few of the females used in these experiments, all of which belonged to the group with fetuses incompatible with respect to H-2 antigens, showed any anti-paternal antibody free in the circulation (Baines *et al.*, 1976).

B-lymphocyte sensitization to paternal antigens is therefore frequent during pregnancy, and detectable levels of circulating antibody against such antigens not unusual.

6.2 Cell-mediated responses to the fetus

The mixed lymphocyte (or mixed leucocyte) reaction (MLR), as a means of assessing lymphocyte response to a cellular antigen, has already been discussed (section 2.6). The intensity of a MLR can be quantified either by measuring the rate of incorporation of radioactively labelled precursors of DNA, or by an assay which estimates concomitant lymphokine release, usually the release of macrophage migration inhibition factor (MIF). An intensive MLR within 24–48 hours of mixing stimulating and responding cells indicates previous sensitization of the latter.

Using this test system, it has been possible to demonstrate MIF production by lymphocytes of pregnant women stimulated by lymphocytes collected from the umbilical cord blood of their babies during the last three months of pregnancy. (Investigation of earlier stages of pregnancy is more difficult, and the risks incurred are less justifiable.) As Figure 6.2 shows, only women in their third or subsequent pregnancies showed a significant degree of cellular sensitization to paternally-derived antigens of their offspring (Rocklin *et al.*, 1973).

Alternatively, multiparity may result in a progressive decline, with successive pregnancies, of the intensity of the MLR, indicating an

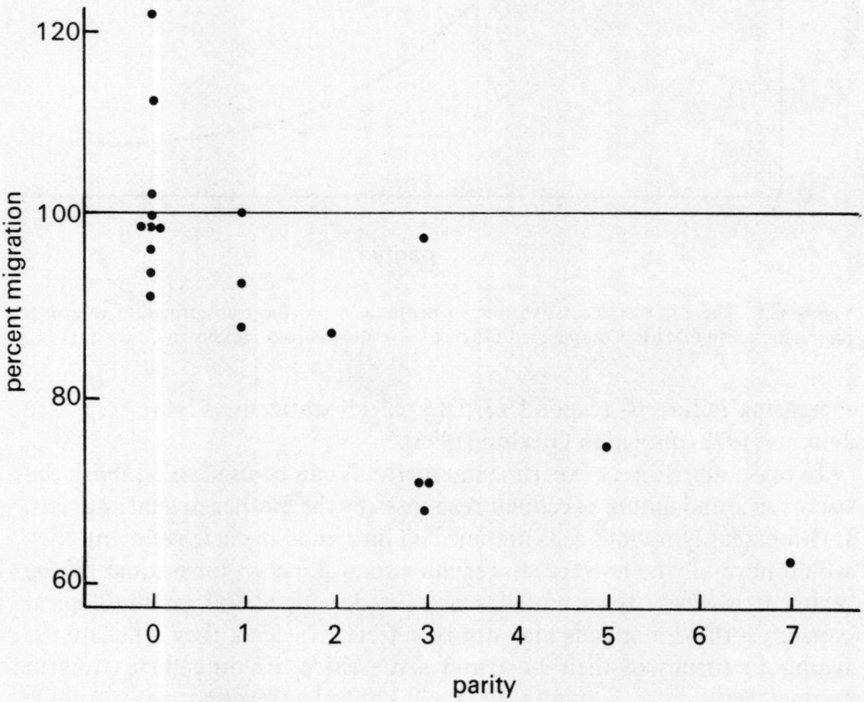

Figure 6.2 Cellular immune response of women to fetal alloantigens. If maternal lymphocytes have become sensitized against alloantigens carried on fetal cells, they respond by producing MIF (p. ●) when confronted with fetal lymphocytes in mixed culture. The degree of inhibition of migration of a standard preparation of (guinea pig) macrophages represents an assay of released MIF, hence of the degree of cellular response (MLR). In this instance MLR intensity increases with the number of pregnancies (based on data in Rocklin *et al.*, 1973).

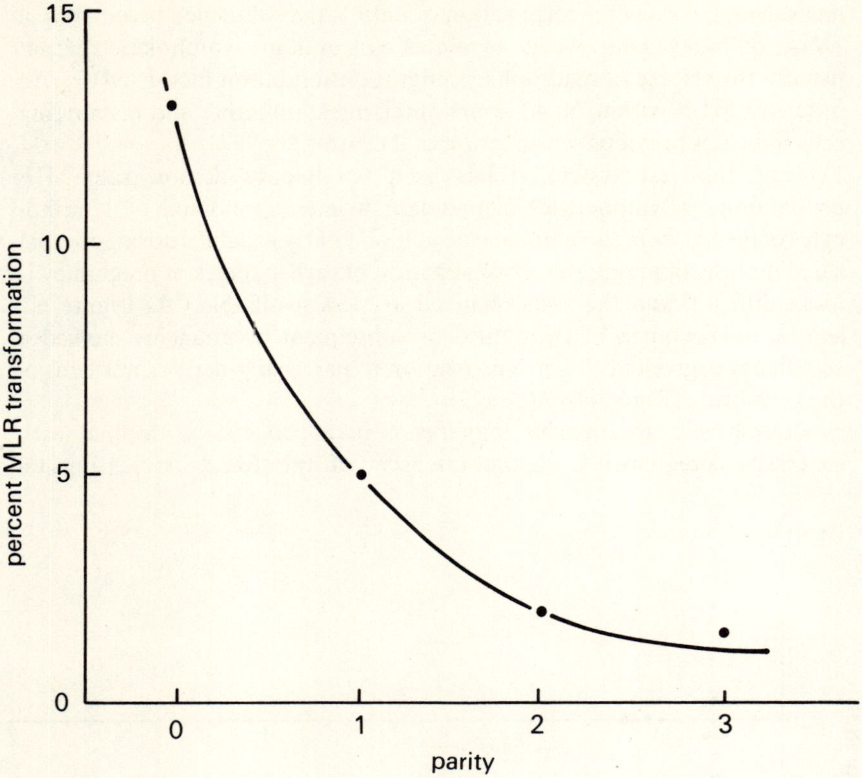

Figure 6.3 The effect of successive human pregnancies on the transformation rate in a paternal/maternal MLR. Compare with Figure 6.2 (redrawn from Jenkins, in Gusdon, 1972).

increasing failure to respond to paternal alloantigens (Figure 6.3)—see Jenkins (1972) quoted in Gusdon (1976).

In mice, more direct experimental methods can be used to establish the occurrence and nature of cellular responses by the mother to fetal antigens. If competent lymphoid cells are injected into an allogeneic recipient, they would normally be rejected. In certain circumstances—for instance, if the recipient is newly born and has not yet developed full immunological capacity—the donor cells may not be rejected. Instead, they infiltrate the lymphoid tissues of their host and settle there. Finding themselves in contact with host alloantigens, they naturally respond with soluble antibody and cell-mediated immune responses; and, despite the gross disparity in size, ungratefully attempt to reject their host. This is known as a graft-versus-host (GVH) reaction. With a neonatal mouse as recipient, the symptoms of a GVH reaction may include inhibition of growth, enlargement of the spleen and severe haemolytic anaemia, and (if donor and recipient differ at the H-2 locus) the result may be the rapid death of the recipient. Donor cells from an individual which has already been sensitized to recipient alloantigens naturally respond more rapidly and intensively than those which become sensitized only after injection.

A GVH presupposes that the recipient cannot, for some reason, recognize the transferred lymphocytes as foreign and reject them before the GVH can develop. If the cells used are from a mouse of one inbred strain, sensitized to antigens characteristic of another, and injected into a hybrid between the two strains, this condition is fulfilled: alloantigens of the injected lymphocytes are not foreign to the hybrid recipient and a GVH may develop. The speed and intensity with which it does will reflect the degree of prior sensitization of the injected cells. The GVH can therefore be assessed quantitatively by measuring the degree of proliferation—either by recording increases in size of relevant regional lymph nodes or spleen, or by measurement of the uptake of radioactively labelled DNA precursors—and this in turn indicates the extent of prior sensitization.

Figure 6.4 shows one set of results obtained by this technique. Female mice of one strain were mated from one to five times with males of another strain (differing only with respect to minor histocompatibility antigens). Within four weeks of birth, lymphocytes were obtained from the females

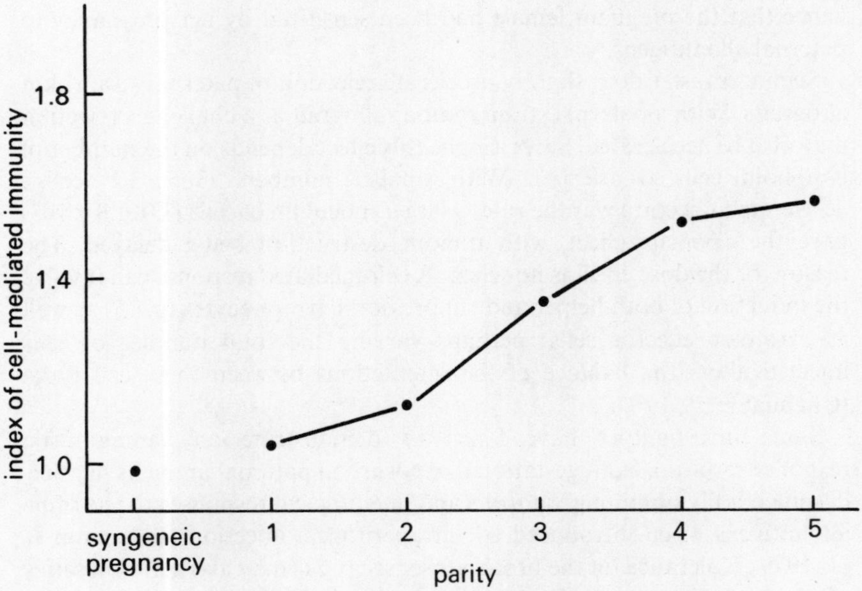

Figure 6.4 Increase in cell-mediated immunity with parity in mice. The technique used is described in the text above; cell-mediated immunity is here expressed as the ratio of the degree of enlargement of the relevant regional lymph nodes in mice injected with cells from pregnant females, compared with that of the lymph nodes of female injected with cells of virgin females of the same strain (based on data from Maroni and Parrott, 1973).

and injected into the footpads of interstrain hybrids. Seven days later, these animals were killed, and the relevant lymph nodes dissected out and weighed. Compared with controls, a significant enlargement of the lymph nodes was found, which was greater with increasing parity (Maroni and Parrott, 1973).

A similar series of experiments, using strains differing at the MHC (H-2) locus as well as combinations differing only at minor loci, and sampling maternal lymphocytes before, during, and at intervals after the fifth pregnancy, produced a somewhat different result. Here cellular immunity was again found, but with the difference that it was only demonstrable in lymphocytes collected from 3 to 6 weeks after birth. The simplest conclusion is that sensitization had occurred while the fetuses were in the uterus, but its manifestations were suppressed until after weaning. An immunosuppressive effect of pregnancy has often been postulated (see section 5.4); an immunosuppressive effect of lactation is a novel concept (Baines *et al.*, 1976).

An alternative method of demonstrating cytotoxic cellular immunity involves transferring lymphoid cells from a female pregnant by a male of a different strain into a recipient of the same strain as herself: accelerated rejection of a paternal strain graft by the recipient in these circumstances would demonstrate that cellular immunity had been passively transferred, hence that the pregnant female had been sensitized by her pregnancy to paternal alloantigens.

Such a transfer does slightly accelerate rejection of paternal strain skin allografts. With a paternal strain tumour allograft as a challenge, rejection may also be accelerated. Surprisingly, this effect depends on the number of lymphoid cells transferred. With smaller numbers ($3\text{–}5 \times 10^7$ cells), accelerated rejection was the rule: a larger inoculum of cells ($1.0\text{–}1.8 \times 10^8$) gave the opposite effect, with tumour destruction being delayed. The reason for the dose effect is not clear. A cell-mediated response can involve the induction of both helper and suppressor T lymphocytes (p. 15) as well as cytotoxic effector cells; perhaps varying the total number of cells injected alters the balance of the interactions between these cell types (Chaouat *et al.*, 1979).

Some investigators have failed to demonstrate any antagonistic response resulting from gestational exposure to paternal antigens in mice, despite readily obtaining vigorous and unequivocal responses to the same alloantigens when introduced by intraperitoneal injection (Wegmann *et al.*, 1979). Tolerance (in the broad sense; see p. 21) may also follow a series of allogeneic pregnancies by the same male, as judged by the acceptance for progressively extended periods of paternal skin or tumour allografts by pregnant mice (Smith and Powell, 1974).

Certain antigens have been described which are either specific to embryonic tissues, or shared by embryonic cells and some adult tumours.

Among these are hormones such as HCG, substances such as α-fetoprotein (AFP) (p. 102), and cellular antigens whose identity is less clearly established (Castro *et al.*, 1974; Gachelin *et al.*, 1976; Lau and Linkins, 1976). Active induction, or passive infusion, of antibody against such 'carcinoembryonic' antigens during pregnancy can lead to abortion (Hamilton *et al.*, 1978; Mizejewski and Grimley, 1976). Spontaneous antibody formation against carcinoembryonic antigens does occur, and although it is not known to impair fertility (Salinas *et al.*, 1978), there may be circumstances in which it does do so.

The immunological consequences of pregnancy are therefore complex, variable, and unpredictable. There is clear evidence that specific cellular immune responses to pregnancy occur in humans and in mice, and that these can take the form either of rejection or of tolerance (Matangkasombut *et al.*, 1979; Saji *et al.*, 1980; Voisin, 1979). In either case, the effect tends to be slow to develop, and can generally be demonstrated, by techniques such as the GVH reaction, only after removal of sensitized lymphocytes from their normal environment. Consequently their relevance, particularly to a first pregnancy, is uncertain. Either the response to fetal alloantigens differs in kind from other comparable immune responses, or interaction between different facets of the immune system in some way explains the anomalies, and results in acceptance of the fetal allograft. It must be added, however, that some workers have failed to find any demonstrable immune effector function developed by pregnant mice against their allogeneic fetuses (Wegmann *et al.* 1979*b*).

6.3 Interactions between cellular and humoral immunity

Both humoral and cellular immune responses to fetal alloantigens occur during pregnancy. Either could prove lethal to the developing fetuses: normally, neither does. General depression of the immune system has already been discussed (p. 93), but as this falls far short of total abrogation of the capacity to respond, it cannot on its own explain fetal survival; nor can the existence of a placental barrier, since this is not complete. Why then is the mother's immune response not fatal?

It is of course an oversimplification to think of the immune system as a unitary, or even a binary system, whose effects are uniformly destructive. The different classes of antibody (p. 12) are not all equally cytotoxic, but have differing properties and actions. The agents of cell-mediated immunity are perhaps less fully understood; but enough is known to indicate a highly heterogeneous population, differentiated with respect to function. Cytotoxic lymphocytes, helper and suppressor cells, and other cell types have already been mentioned (section 2.3). Since immune effectors react to foreign antigenic sites, there will be a measure of competition for these sites, with the possibility of the more cytotoxic effectors being competitively excluded. By manipulating the balance between different components of the immune system, a fetus might contrive to ensure its own safety. Reducing the impact of the mother's immune response in this way would be specific with respect to the alloantigens of the fetus, in contrast to the suppression induced by substances such as AFP and other glycoproteins, or by nonspecifically acting suppressor T cells.

Evidence that such mechanisms operate in relation to the survival of tumours has been accumulating for many years. Following early attempts to preimmunize recipients of tumour grafts against tumour-specific antigens, it was found that growth of the grafts was enhanced rather than diminished or abolished. This effect was attributed to the induction of harmless antibody which then impeded the destructive cell-mediated immune response to the tumour (Currie, 1974).

As far as the mammalian fetus is concerned, it has been shown that lymphoid cells from allogeneically pregnant female mice inhibit the growth in culture of cells derived from embryos of the paternal strain. This cell-mediated inhibitory effect can be blocked by the addition to the medium of blood serum from the pregnant mice. The serum of syngeneically pregnant females—that is, females pregnant by males of the same genetic strain as themselves—does not block the growth of cultured embryonic cells; the effect is therefore specific and not a general result of pregnancy. Further experiments have shown that lymphocyte-mediated cytotoxicity is also specifically blocked by serum from appropriate pregnant females (Hellström *et al.*, 1971).

Similar 'blocking factors' appear in the blood serum of pregnant women. As already mentioned, the mixed lymphocyte reaction (MLR) can be used to assess the degree to which the responding cells have previously been sensitized to alloantigens of the stimulating cells, while lymphokine production is a correlate of the MLR which can be exploited to measure its intensity. MIF is the lymphokine whose release can most easily be assayed (p. 24). If a MLR is carried out between lymphocytes of a pregnant woman as responders, and paternal lymphocytes as stimulators, MIF release shows that the fetus has sensitized the mother's lymphocytes to paternal alloantigens. When the blood plasma of the pregnant woman (and, in general, not that of other pregnant women) is added to the culture medium, MIF release is reduced; the response must therefore be blocked by a specific factor in solution in the plasma. Similar experiments have shown blocking activity is present also in amniotic fluid (Gleicher *et al.*, 1980; Pence *et al.*, 1975).

Blocking activity has been reported to be deficient in the blood plasma of women with a history of recurrent spontaneous abortions; in one case, of a woman who had aborted three times in succession, a successful fourth pregnancy was accompanied by the development of blocking capacity in the blood plasma. Some forms of spontaneous abortion might therefore be precipitated by a failure of the normal mechanisms which protect the fetus (Rocklin *et al.*, 1976).

Three candidates for the role of specific blocking factor have been proposed: antibody; fetal alloantigen; and immune complexes incorporating antigen, antibody, and probably components of complement.

Evidence is strong that specific antibodies are involved. Blocking activity in multiparous women is associated with the IgG fraction of blood serum, and it can be removed from the serum by absorption with paternal lymphocytes. Several possible mechanisms have been advanced to explain the protective action of specific antibody. It might mask antigenic sites on the surface of fetal cells by combining with them and preventing their recognition and destruction by helper or cytotoxic effector cells. This view is supported to a certain extent by the finding, discussed on p. 114, that HLA antigens of maternal origin are readily demonstrated on fetal lymphocytes, but paternal alloantigens are often more difficult to demonstrate. When the mother has circulating cytotoxic antibody against paternal alloantigens, her serum fails to kill fetal cells. After a period in culture, this is no longer the case: paternal antigens reappear and the fetal cells again become vulnerable to specific cytotoxic sera. The simplest (although not the only) explanation of this is that incubation allows time for the removal of masking antibody, or for the expression of fresh paternal antigen (Bernard, 1977; Tiilikainen *et al.*, 1974).

For masking to succeed, the masking antibody would have not to be itself cytotoxic. This requirement would be fulfilled if the majority of the attaching antibody molecules were not of a cytotoxic type; recently it has been shown that the predominant antibody produced during pregnancy in the mouse in response to fetal antigens does in fact belong to a non-complement-fixing and non-cytotoxic subclass, IgGl. There is also some evidence that IgG production increases during pregnancy relative to IgM. IgM is strongly complement-fixing, but enters the fetus in negligible amounts. Of the subclasses of IgG, non-complement-fixing IgG3 seems to be transported into the fetus 20–50 times more rapidly than IgG of other subclasses. These observations are all compatible with the concept of masking of antigenic sites by antibody which is itself relatively harmless (Bell and Billington, 1980; Ralph *et al.*, 1972).

Even if the antibody attaching to fetal antigenic sites was not itself intrinsically non-cytotoxic, it might still fail to destroy target cells if complement was absent or deficient, or if the antigenic determinants exposed on the fetal cell surfaces were too thinly dispersed for complement assembly to proceed. The question of complement deficiency in fetuses has already been discussed (p. 114), while the apparently low antigenicity of placental tissues, and of skin allografts from neonates (p. 109) might also indicate a sparseness of antigenic sites compared with adult tissues (Linscott, 1970).

Masking of target antigen is one possible explanation for failure of effector cells to act. An alternative explanation is that the antigenic molecules might actually be stripped from the surface of the fetal cells, or internalized, following their combination with antibody. If the disappearing antigen molecules were not immediately replaced by fresh synthesis, either mechanism might decrease antigenicity of the cells with respect to paternal alloantigens. In practice, this would be hard to distinguish from the effects of masking. Modulation of the antigen phenotype of tumour cells in this way is well established and in some circumstances can persist for a significant period of time; a similar mechanism might make some contribution to fetal survival.

The case for efferent blocking (or enhancement) by interference with the recognition of target antigen by sensitized effector cells is a strong one. Also possible is an afferent action, with specific antibody preventing sensitization of the cell-mediated immune system in the first place. Masking of Ia-like antigenic sites (in humans, products of HLA-D loci; see p. 20) by antibody would impede cellular recognitions and interactions necessary for the induction of helper T cells, hence of cytotoxic lymphocytes (p. 100). Presentation of antigen such as HLA-A or HLA-B without the concomitant presentation of Ia antigenic determinants, is thought to predispose to tolerance rather than immunity. Antibodies have been detected, in the sera of pregnant women, which react with appropriate B cells, but not with T cells. As Ia antigens (or their human equivalents) are carried by B cells but are largely absent from T cells, this is suggestive evidence for the formation of anti-Ia antibody during human pregnancy (Winchester *et al.*, 1975).

Blocking ability does not always correlate with the serum content of relevant antibody. This does not invalidate the concept of antibody having blocking action: if the action involves adsorption to antigenic sites, clearance of blocking antibody from the circulation might be rapid. Direct evidence of B-cell activity during pregnancy might therefore prove more relevant, and this evidence has already been mentioned (p. 117). A finding possibility more telling against the necessary involvement of humoral antibody is that successful pregnancy is possible in women affected by agammaglobulinaemia, that is, with a genetically-determined defective capacity to synthesize humoral antibody. If blocking antibody was essential for fetal survival, such pregnancies would be impossible (Loke, 1978).

Specific blocking factors need not, however, be antibodies. Lymphocytes from patients with advanced cancer must be thoroughly washed before they can be effectively cytotoxic. Their cytotoxicity is abolished again by addition of the patient's serum. A specific blocking factor is present with an affinity for lymphocytes rather than for their target cells; this makes it more likely to be antigen than antibody, a suggestion which is supported by the finding of blocking factor in the sera of some cancer patients of molecular weight too low, at 30 000, to be immunoglobulin (Currie, 1974).

There is no good evidence for blocking antigen in pregnancy. Soluble histocompatibility antigen does not appear in the circulation in large—or even, generally, detectable—amounts. However, in the immediate vicinity of the fetus this might not be the case. Histocompatibility antigens on fetal cells are more readily detached (by, for example, mild sonication) than those of adult cells. Possibly sufficient antigen is shed in and around the fetus to afford some protection, in much the same way as aircraft can scatter large amounts of metal foil strips to conceal their own radar image (Alexander, 1974).

Finally, immune complexes of antigen, antibody and complement components may have an inhibitory role, by acting centrally on lymphocytes. This operates in some cases of the enhancement of tumour growth. Some investigators report a rise in the level of circulating immune complexes during normal human pregnancy, although others find this only in abnormal pregnancies and regard it as a pathological symptom. The ability to phagocytose immune complexes appears also to be stimulated by pregnancy (Masson *et al.*, 1977; Pachi *et al.*, 1980; Rosenthal, 1977).

Specific blocking of immune responses by antibody, and its significance in pregnancy, therefore remains something of a mystery. Several possible agents are present, and several possible mechanisms proposed for their action, based to a large extent on models of tumour growth. These agents and mechanisms are not mutually exclusive: their relative importance remains to be elucidated.

6.4 Suppressor cells

The acceptance of paternal allografts following multiple pregnancies (p. 122) might have one of several causes. One possibility is that lymphocyte clones against paternal antigens are eliminated or depleted by constant exposure to antigen, in a 'clonal deletion' form of immune tolerance (section 2.5). Female mice of some strains can become tolerant to the male-specific antigen H-Y, after repeated syngeneic pregnancies. The transfer of T lymphocytes from the spleens of animals tolerant to H-Y into virgin females renders the recipients tolerant in their turn. This does not accord with a clonal deletion explanation for the tolerance. The transferred cells must include some with the ability to suppress a specific immune response of the recipient, the cell-mediated rejection of cells bearing H-Y antigen (Smith and Powell, 1974).

Passive transfer of tolerance can also be achieved with allogeneically pregnant mice where male and female differ at the major histocompatibility locus H-2, not just in respect of minor antigens such as H-Y. In this case, spleen cells strongly suppress the response of maternal to paternal strain cells in the mixed lymphocyte reaction test. The suppressive capacity is eliminated by adding to the suppressor cell population of spleen cells both complement and a complement-fixing antiserum against the Thy-1 antigen characteristic of T cells (p. 14). This establishes that the suppression is a property of specific suppressor T cells, and not by the production of blocking or enhancing antibody by B cells sensitized before transfer (Baines *et al.*, 1980; Chaouat and Voisin, 1979, 1980).

Although some suppressor cells are found in the spleen, a major site of suppressor cell generation appears to be within the regional lymph nodes which drain the uterus. Cells in the draining lymph nodes of pregnant mice appear to act by suppressing the development of cytotoxic lymphocytes from their precursors, and to do so non-specifically with respect to paternal antigens. This action appears to be effected by a soluble material, of which little is known other than that its molecular weight is greater than 25 000 (Clark and McDermott, 1978; Clark *et al.*, 1980; Gottesman and Stutman, 1978).

The relation between nonspecific and specific suppressor T cells is unclear; however, it seems most unlikely that in either case the antigens to which the response will be suppressed by the uterine lymph nodes will be fetal ones. The importance of the induction of suppressor cells is suggested by the finding that female mice of a certain strain which spontaneously abort their offspring during pregnancy have reduced suppressive capacity and greater cytotoxic lymphocyte activity than mice of the same strain with normal and uneventful pregnancies (Clark *et al.*, 1980).

The draining lymph nodes of the uterus appear to be implicated in local immunosuppression in rats also, since their excision before conception permits immune responses to paternal strain alloantigens to develop during pregnancy when they would not otherwise do so (Beer and Billingham, 1978).

The lymph nodes which drain the uterus may therefore have an important role in locally generating immunoregulatory suppressor T cells. Possibly suppressor cell generation here is facilitated by immunosuppressive agents emanating from the fetus; it seems reasonable to suppose that substances such as AFP (see p. 102), known to induce suppressor cell formation, might be present in higher concentrations in the lymph nodes draining the uterus than in more distant elements of the lymphoid system. Additionally, any maternal lymphocytes which cross the placenta into the fetus will be exposed to high levels of AFP, and if they then re-emerge from the fetus, it is likely that they will settle, in the first instance, in these same lymph nodes.

There are obvious advantages in a pregnant female limiting her tolerance of alien antigens as far as possible to the vicinity of the uterus. However, signs of immunological response to pregnancy are detectable

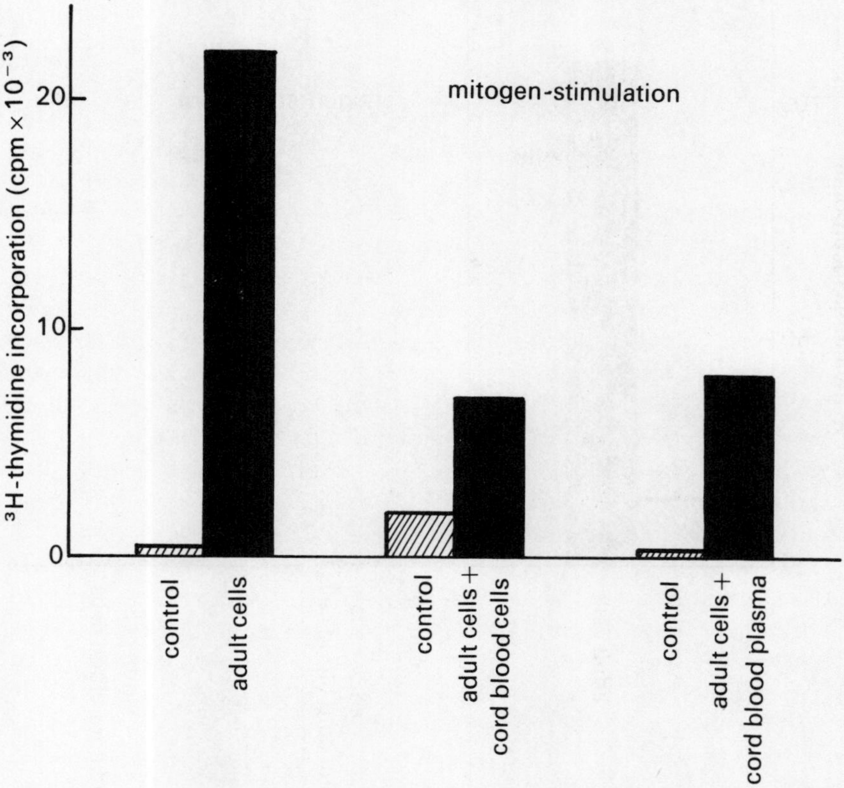

Figure 6.5 Suppressive effect on adult lymphocyte proliferation of leucocytes collected from the umbilical cord blood of human infants, pre-term and immediately after birth.

(a) Con A-induced proliferation of adult lymphocytes is suppressed in the presence of cord blood leucocytes or cord blood plasma. The left hand column of each pair refers to the proliferation of adult lymphocytes in the absence of mitogen.

elsewhere in the body, while excision of the draining lymph nodes does not prevent (or even impede) conception, so the importance of this limitation is likely to be slight.

Suppressor cells may be generated by the lymphoid system of the fetus, as well as by that of the mother. Lymphocytes collected from the umbilical cord blood of healthy human infants have a potent suppressive effect on the proliferation of adult lymphocytes. Figure 6.5 shows the results of stimulation by the mitogen Con A, and by an antigenic extract of *Candida albicans*, a fungus which commonly infects the genital tract. In each case the suppression by cord blood cells is marked (Abedin and Kirkpatrick, 1980).

The fetal cells responsible appear to be T lymphocytes which act, in part at least, by the release of a soluble factor of some sort, since cord blood

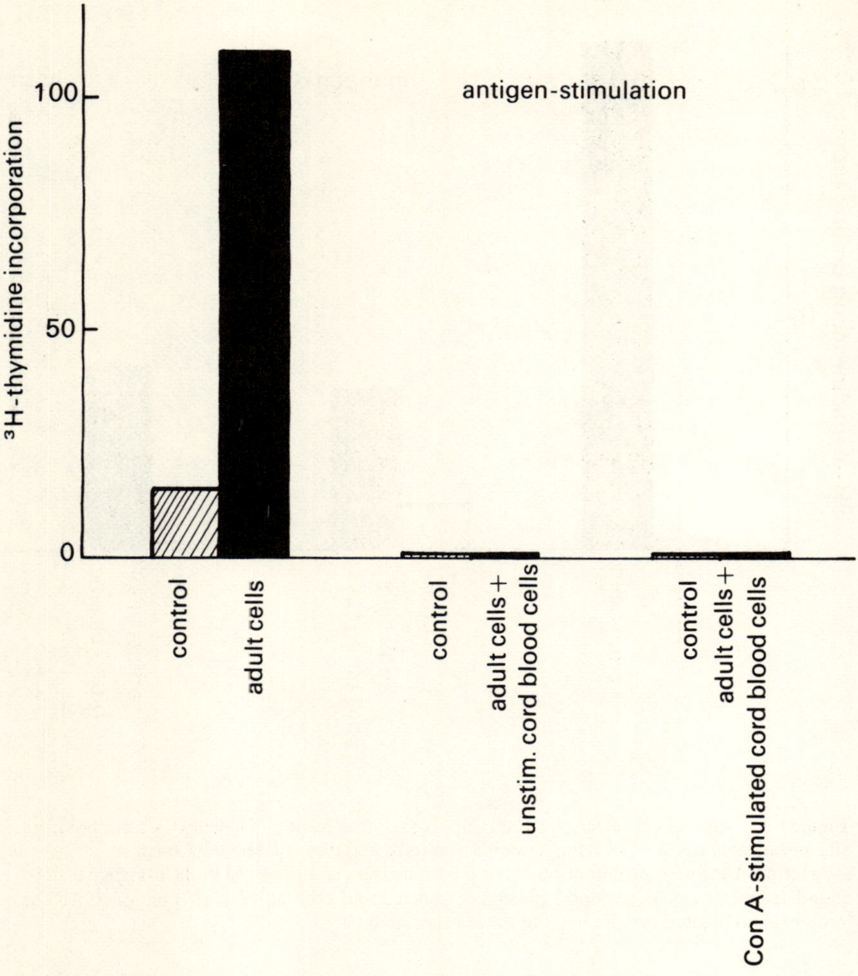

Figure 6.5 (b) Antigen-induced proliferation of adult lymphocytes is suppressed by cord blood leucocytes, whether or not the fetal cells are themselves first stimulated to divide with mitogen (Con A). This contrasts with adult suppressor cells where suppression is conditional on cell division. Left hand columns represent unstimulated adult lymphocytes.

Proliferation in each case is assessed as the uptake of labelled DNA precursor. (Based on data from Abedin and Kirkpatrick, 1980).

plasma and the culture medium in which fetal T cells have been kept, are as effective as the cells themselves.

In the experiments described, the fetal cells did not themselves require to proliferate in order to suppress. This conflicts with the conclusions of other workers, who have inferred that proliferation of the fetal cells—achieved *in vitro* by PHA stimulation—is indeed necessary for the release of the soluble suppressive factor. There have also been reports of fetal cells rather than lymphocytes with suppressive action; phagocytic monocyte-like cells have been implicated.

Fetal suppressor cells might be important in suppressing the activity of marauding maternal cells. Maternal lymphocytes do occasionally cross the placenta, enter the fetal circulation, and even settle in fetal lymphoid tissues. Naturally occurring symptoms of a fetal or neonatal graft-versus-host reaction are extremely rare (p. 115); it is possible that the rarity is due to suppression of the potentially aggressive maternal cells. Moreover, newborn rabbits are much more resistant to the artificial induction of a GVH reaction (by injecting competent adult lymphocytes) than adult rabbits whose immune systems have been weakened by X-irradiation. Again, fetal suppressor cells may suppress the interlopers.

An alternative effect of the fetal suppressor cells may be to prevent a fetal response against maternal alloantigens. In a typical outbred population, with extensive heterozygosity, many alloantigens carried by mothers will not be present on their fetuses, and might therefore elicit a fetal immune response. The fetal immune system is, in humans at least, quite capable of responding to antigen before birth in the MLR but is, for the most part, incapable of proceeding to full effector function, as judged by the criterion of cell-mediated lympholysis (Rayfield *et al.*, 1980).

6.5 Further interactions between mother and fetus: the importance of being different

The occurrence of blocking antibody and suppressor cells shows that a maternal immune response to the fetus need not be harmful, and in some circumstances might positively benefit it. There have been numerous reports of further supposed beneficial effects of antigenic disparity between mother and fetus, including increased placental weight, improved fetal growth rates, and enhanced intrauterine survival prospects. These effects were attributed to some form of immune response to fetal antigens, more intense (hence more beneficial) the more different the fetal alloantigen complement was from that of the mother. The beneficial effects appeared to be further enhanced by prior immunization of the mother to fetal (or paternal) antigens, and diminished by the induction of specific immune tolerance to those antigens (Beer and Billingham, 1976; Hamilton and Hellström, 1978).

The advantages in being different from one's mother could explain the remarkable degree of polymorphism of major histocompatibility antigens. Unfortunately for an elegant hypothesis, the design and interpretation of these experiments has been severely criticized on various grounds and they can, for the most part, be discounted (McLaren, 1975).

Some residual observations, however, still suggest that antigenic disparity may play a role other than a harmful one in materno-fetal relations.

In humans, for instance, there appears to be a slight correlation between placental weight and the intensity with which maternal lymphocytes respond to fetal ones in the MLR, although not all investigators have found this (Jenkins and Good, 1972).

A further example appears, at least in the restricted context of crosses within highly inbred strains of laboratory rat, to be of considerable importance. When a backcross is carried out between an interstrain hybrid male rat and a female of the same inbred strain as its mother, approximately half of the offspring should be homozygous and half heterozygous with respect to any genetic loci at which the initial strains differed; inbred strains being presumed homozygous (Figure 6.6). The same is of course true of the reciprocal cross, between an interstrain hybrid female and a paternal strain male; genetically, these two crosses are symmetrical.

If we consider a gene specifying a histocompatibility antigen, then the two crosses are no longer interchangeable, since in the first case half of the progeny will be antigenically incompatible with their mother, while in the latter all will be compatible.

If this is carried out using two parental strains differing at the major

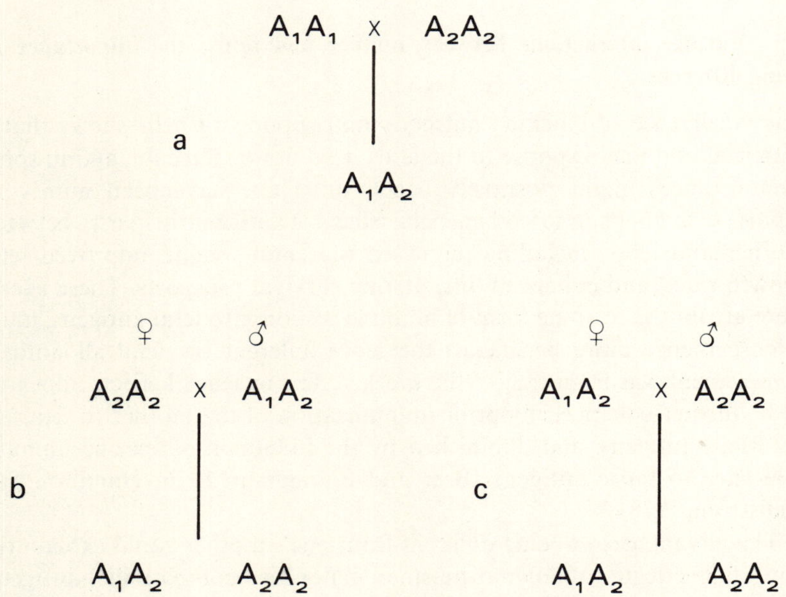

Figure 6.6 (a) represents the initial cross between members of two genetic strains, each of which is homozygous with respect to different alleles of a histocompatibility locus A. All offspring are heterozygous A_1A_2. In crosses (b) and (c), both homozygotes (A_2A_2) and heterozygotes (A_1A_2) are produced. In the former case, the heterozygotes are incompatible with their mother by virtue of the A_1 allele, while in the latter they are compatible.

histocompatibility locus (and also, probably, at an indeterminate number of unlinked minor antigenic loci), the progeny are found in practice to include a significant excess of heterozygotes in the cross where these are *incompatible* to their mother with respect to MHC antigens. Where heterozygotes and homozygotes are both MHC-compatible to their mother, they appear in roughly equal numbers as predicted.

How does this come about? Either heterozygous conceptions are favoured in an incompatible mother, or homozygotes compatible with respect to MHC antigens (but incompatible at the minor loci) are discriminated against during pregnancy. The latter explanation appears to be the correct one. Progeny compatible with their mothers at the MHC locus do appear to have a higher mortality rate, both before and after birth.

Many of those which die soon after birth show symptoms of a 'wasting syndrome' characteristic of a graft-versus-host attack (p. 120): a hunched posture, ruffled fur, low body weight, and abnormalities of the lymphoid organs. This suggests that the infant mortality—and, by extension, prenatal mortality as well—might be caused by infiltration of maternal lymphocytes into the fetus. The fetus in the particular genetic situation established by the backcross will fail to recognize these as foreign, and so will accept them, since maternal cells carry no antigenic determinants foreign to the fetus. On the other hand the fetus will carry minor histocompatibility antigens foreign to the mother, and the maternal lymphocytes which have crossed the placenta will address their attention to these. The result is a GVH reaction.

The corollary is that in crosses where mother and fetus differ in MHC genotype, infiltrating maternal cells will be recognized and disposed of by the fetal immune system before causing any damage. The extreme polymorphism of the MHC system will ensure that in normally outbred populations MHC compatibility between mother and fetus will be exceptional; the prevailing incompatibility at the MHC locus will therefore serve to protect the fetus against potential GVH attack on account of differences at minor histocompatibility loci (Palm, 1974).

In humans, pregnancy may also be adversely affected by a degree of MHC compatibility. Appropriate genetic crosses cannot be arranged, for experimental purposes, in advance. Instead of arranging a particular cross and looking for abnormalities of reproduction, the experiment is, as it were, carried out retrospectively. Women with a history of disorders of pregnancy are investigated for an unusual degree of MHC compatibility with their husbands. A greater degree of parental HLA compatibility than expected has in fact been found in couples who have had either several spontaneous abortions, or a high incidence of neural tube defects among their offspring (Table 6.1), although there is no direct evidence to implicate a GVH reaction by infiltrating lymphocytes (Shacter *et al.*, 1979).

A form of wasting syndrome does occur in human infants, but is

Table 6.1 Degree of sharing of antigens of HLA-A and HLA-B loci by the parents of children born with neural tube defects, or with a history of spontaneous abortion (data from Shacter *et al.*, 1979)

No. of pregnancies	*No. of spontaneous abortions*	*Infants with n.t. defects*	*Couples with 2 or more antigens in common (%)*
3+	0	0	0
2+	1	0	6
3+	2+	0	22
1+	0	1+	35

exceedingly rare, while it has also been tentatively suggested that certain diseases affecting the lymphoid system of adults, such as Hodgkin's disease, might stem from the activity of latent maternal lymphocytes. In neither case has a GVH reaction been directly established as the cause, nor an unusual degree of MHC compatibility identified.

If mortality in the progeny is limited by the vigilance of the fetal immune system, it follows that a fetus must acquire immune competence before birth. In the human fetus, this is certainly the case. Small lymphocytes appear in the peripheral blood at about seven weeks from conception. By ten weeks or so, T lymphocytes are competent to respond to mitogens and in the mixed lymphocyte reaction, and by about the same age B cell competence is shown by the synthesis of both IgG and IgM. There is no reason to believe that the system is radically different in species like the mouse or rat, whose gestation periods are brief in duration. Fetal immune systems appear competent well before birth, although the surrounding presence of the mother will greatly limit fetal experience of antigens (other than maternal ones)— see Jones (1976); Solomon and Horton (1977).

Among the different classes of T lymphocytes, the presence of fetal suppressor cells is now well established in newborn human infants and late fetuses, as well as in fetal mice; in the latter case, the high levels of AFP on the fetal side of the placenta is conducive to their formation (pp. 102 and 128). Perhaps the fetal suppressor cells help to hold in check any maternal cells which cross the placenta, until they either pass back into the mother or can be destroyed as the fetal immune system becomes more aggressive after birth (Baines *et al.*, 1976; Chaouat *et al.*, 1979; Murgita *et al.*, 1977; Oldstone *et al.*, 1977).

A fully competent immune system could prove an embarrassment to the mother, if fetal lymphocytes could enter the maternal circulation and develop a GVH against maternal tissues (the exact reverse of the situation discussed earlier in this section). There is evidence that this can occur, as indicated by enlargement of the para-aortic lymph nodes (p. 90) in rats in crosses where the mother is antigenically foreign to the fetus, but not *vice versa*. If this lymph node enlargement is indeed a sign of a fetus-versus-mother GVH reaction, it appears to go no further than the local lymph nodes of the uterus (Beer and Billingham, 1974*a*).

A further interaction between the immune systems of mother and fetus is in the control of fetal immunoglobulin synthesis. The large amounts of maternal antibody which enters the fetus, particularly in late pregnancy (see p. 113 and Chapter 7), inhibit antibody production by the fetal immune system. Not only is the total amount of fetal immunoglobulin reduced, but the type of molecule may also be regulated. Immunoglobulins show limited polymorphism of the constant region as well as the extreme variability of the antigen-combining variable region (Chapter 2). If, for example, a female rabbit is homozygous for one allele at a locus determining one such constant region allotype, and mates with a male homozygous for a different allele, the heterozygous offspring should express both of these allotypes, and in roughly equal amounts. This is, of course, independent of the specificity of the antigen-binding regions of the immunoglobulin in question.

As far as the female is concerned, one of these allotypes (considered as an antigen) is foreign. If, in such a cross, the female is immunized against the paternal allotype, the result may be partial suppression of synthesis by the fetus of immunoglobulin bearing that allotype, and compensatory production of the maternal allotype. This situation may persist after birth. Anti-allotype antibody may arise naturally: this has been reported in human pregnancies (Catty and Lowe, 1976; Fudenberg and Fudenberg, 1964; Leiper and Solomon, 1976).

Mother and fetus thus interact immunologically in many complex and subtle ways which, although imperfectly understood, are crucial to maternal acceptance of the fetal allograft during pregnancy, To consider only this point—the fetus as an allograft, and why the mother fails to reject it—is probably to take an excessively narrow and restricted view of the immunological relationship between mother and fetus: there is evidently much more to it than that.

6.6 The role of the placenta

In an earlier section (p. 108) the antigenicity of trophoblast was considered, and compared with that of the inner cell mass of the blastocyst and its derivatives. Major histocompatibility antigens appear to be expressed on the trophoblast of preimplantation embryos only briefly and at low levels; around the time of implantation surface H-2 on the conceptus declines to undetectable levels. Minor histocompatibility antigens may continue to be expressed on trophoblast, while the inner cell mass probably expresses a normal complement of antigen.

As implantation proceeds and the definitive placenta is established, the trophoblast layer develops and differentiates in structure and function.

Apart from its nutritive role, it elaborates hormones and other substances, some of which may make a significant contribution to fetal survival by acting as immunosuppressants, either locally or throughout the pregnant female (p. 98). Does the trophoblast of an established pregnancy still serve as an immunological barrier between mother and fetus, of sufficiently low antigenicity itself not to provoke a maternal immune response?

Most investigations of trophoblast antigenicity involve the culture of trophoblast explants *in vitro*; histocompatibility antigens, if present, can then be detected serologically by the use of specific antisera. The use of artificially cultured cells, although convenient, results in problems of interpretation. Firstly, trophoblast is a heterogeneous tissue; disrupting and mixing trophoblast cells for *in vitro* study inevitably makes it difficult to distinguish the different cell types, to tell which are antigenic and which not, and what their juxtapositions are in an undisturbed conceptus. Secondly, antigens may not be uniformly distributed over the surface of a cell. In its normal position, a trophoblast cell may carry antigenic determinants, but partition these on the cell surface so that the face towards the uterine endometrium is not antigenic. A situation similar to this has been demonstrated in the yolk sac endoderm of mice (p. 109)—see Parr *et al.* (1980). Thirdly, explanted cells may respond to their new environment by synthesizing antigens which they would not normally carry; and, finally, an apparent lack of antigenicity inferred by one method does not mean that other, more sensitive methods might not lead to a different conclusion. Caution is therefore needed in interpreting evidence of trophoblast antigenicity based predominantly on the investigation of trophoblast cells taken out of their normal context.

The major layers of trophoblast found in the mouse placenta are illustrated diagrammatically in Figure 5.3 (p. 86); a thin outermost layer of giant cells, beneath which is a thicker layer of spongiotrophoblast, surrounding a core of labyrinthine trophoblast. All three major layers are in direct contact with maternal tissues, the first two abutting the decidual tissues of the endometrium, and the labyrinthine trophoblast—or at least the outermost of the three layers into which it can, in turn, be subdivided—is bathed by maternal blood (Billington, 1975).

It is now established that trophoblast giant cells fail to express MHC antigens. The giant cell layer might therefore still serve to impede recognition of an allogeneic conceptus by its mother, although in view of other materno–fetal contacts this cannot be the sole means of protection. Labyrinthine trophoblast is probably also lacking in histocompatibility antigens of the MHC. When dispersed *in vitro*, a small percentage of its component cells do react with antisera specific for (paternal) H-2 antigen, but these MHC-positive cells may be contaminant cells of the fetal mesenchyme. The bulk of labyrinthine trophoblast apparently carries no H-2 antigens (Jenkinson and Owen, 1980; Sellens *et al.*, 1978).

MHC antigens might in fact be present, but concealed under some form of masking layer. It has been suggested that the low antigenicity of trophoblast cells is due to an overlying coat, or glycocalyx, of sialomucin material. This might conceal antigenic sites, or prevent attachment by passing lymphocytes which would in consequence fail to become sensitized to trophoblast antigens. Treatment of cultured trophoblast cells with the enzyme neuraminidase has been reported to strip the sialomucin coat and to increase immunogenicity. Not all investigators, however, have confirmed this, and there are in any case alternative ways in which neuraminidase might potentiate an immune response besides unveiling hidden antigens (Bagshawe and Lawler, 1975; Martin *et al.*, 1974; Wynn, 1969). It is likely, therefore, that the low antigenicity of certain forms of trophoblast is due simply to a failure of expression of antigens rather than to their concealment.

In contrast to the giant cell layer and the labyrinthine trophoblast, the spongiotrophoblast layer is antigenic (although it has not conclusively been shown that the antigens are exposed *in vivo*). Antisera against

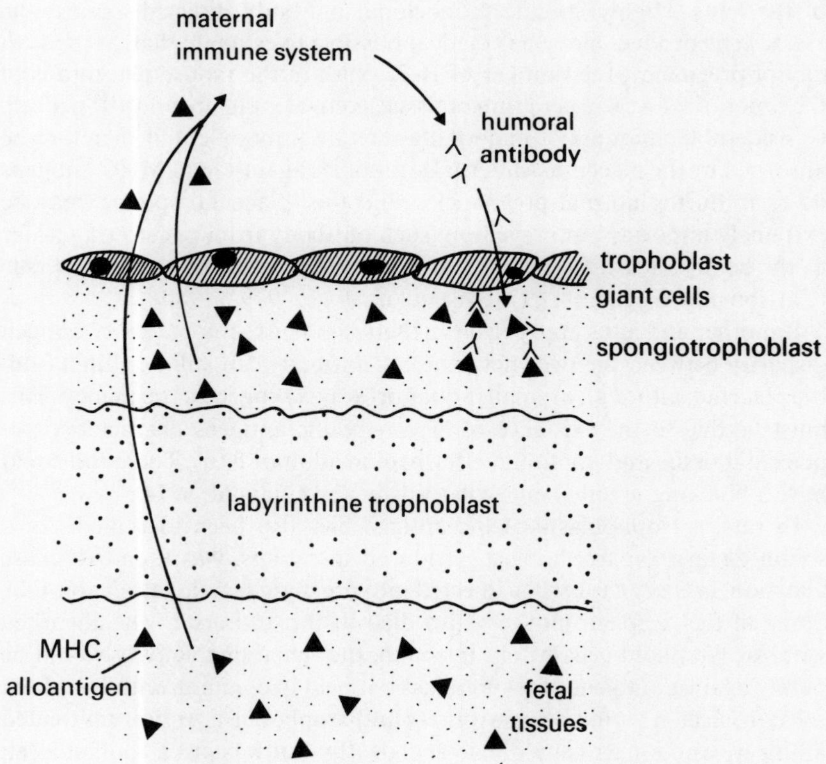

Figure 6.7 Trapping of antifetal antibody by the mouse placenta (p. 138).

particular products of the MHC locus have made it clear that the antigens expressed are almost certainly those determined by the K and D loci (p. 19). Ia antigens are absent from trophoblast of all stages of the mouse embryo, and from embryonic tissues as well. Ia antigens are, however, strongly expressed on maternal decidual cells, and it is possible that these Ia sites play some part in presenting, to the maternal immune system, H-2D and H-2K antigenic material shed from the spongiotrophoblast or from the embryo itself (Jenkinson and Searle, 1979).

Evidence of placental antigenicity in undisturbed pregnancies comes from the finding that maternal antibody, specific to paternal strain alloantigens, is bound to placental tissue and can be eluted from there. Mostly, this is IgGl and is not complement-fixing and therefore not cytotoxic; it has blocking or enhancing properties and can, for instance, be used to prevent mice of the maternal strain from rejecting tumour grafts of paternal strain origin. It may be similarly protective to the placental allograft (Voisin, 1979).

An alternative view is that absorption by the placenta of maternal antibody against paternal strain alloantigens is essential for the well-being of the fetus. Highly specific monoclonal antibody directed against the H-2K gene product alone has made it possible to estimate that, at the 17th day of pregnancy, the number of H-2K sites in the mouse placenta is of the order of 1.1×10^{13} per gram of tissue, equivalent to around 10^4 per cell. Considerable amounts of maternal antifetal antibody could therefore be absorbed by the placenta. Maternal antibodies against fetal MHC antigens do form during normal pregnancies, and this 'placental sponge' may be extremely important in preventing such antibody from penetrating as far as the embryo itself. Figure 6.7 illustrates the possible sequence of events (Carlson and Wegmann, 1978; Wegmann *et al.*, 1979*a*, 1980).

If mother and fetus are syngeneic, there is no histocompatibility antigen disparity between the two; nevertheless, antibody can still be eluted from the placenta, although in smaller quantities, in a syngeneic pregnancy. This must be due to the existence of organ-specific antigens characteristic of placental tissue and not found elsewhere in adult or fetus. These too could have a blocking or enhancing role (Voisin and Chaouat, 1974).

In rats, a trophoblast-specific antigen has also been identified. Antiserum against rat trophoblast, produced in rabbits, was found to cause abortion, in every case, when injected into pregnant female rats (Table 6.2). Some of the recipient mothers also died. If the antiserum was absorbed with rat lymphoid cells before injection, the remaining activity would be purely against antigens of trophoblast not held in common with lymphoid cells; in effect, it would become purely anti-trophoblast. Antiserum treated in this way no longer caused maternal deaths, but was just as potent as an abortifacient, and was equally effective in all strains of rat tested. This indicates a placenta-specific antigen (or antigens) common to all strains of

Table 6.2 Effect of infusing pregnant female rats with antisera, prepared in rabbits, against trophoblast; the antisera were absorbed with various cell types to remove defined categories of antibody before infusion

		Number of females		
Antisera against	*Absorbed with*	*Treated*	*Aborting*	*Dead*
—*	—	8	0	0
trophoblast	—	12	12	2
trophoblast	lymphocytes	12	12	0
lymphocytes	—	12	3	8
trophoblast	trophoblast +lymphocytes	10	0	0

* Unimmunized control.

rat, and not found in tissues other than placenta. It also suggests the possibility that some otherwise unexplained spontaneous abortions in humans might be caused by a maternal immune response to placenta-specific antigens; and that deliberately immunizing against such antigens might ultimately provide the basis of a form of artificial contraception (Beer *et al.*, 1972).

Much less is known of the antigenic status of trophoblast in the early stages of human pregnancy; availability of material for study is restricted to full term placentae, and the placentae of clinically aborted fetuses (Faulk, 1980; Faulk and Johnson, 1977; Faulk *et al.*, 1978; Johnson *et al.*, 1980; Whyte and Loke, 1979).

Most evidence favours the view that antigens of the major histocompatibility complex are not expressed on trophoblast cells of an established human placenta, although some HLA antigen may be detectable on non-trophoblast cells within the placenta. This applies to antigens corresponding to the Ia antigens of mice (HLA-D and DR) as well as to the classical serologically-defined MHC antigens (HLA-A, -B and -C). The human amnion similarly lacks HLA antigens. Presumably this will reduce the likelihood of the mother becoming effectively sensitized to paternal alloantigens at the placental interface (Barnstaple and Bodmer, 1978; Faulk and Temple, 1976; Faulk *et al.*, 1976; Sundqvist *et al.*, 1977).

There is, however, some evidence that trophoblast may not be totally lacking in HLA antigens, or in the capacity to induce an immune response. Hydatidiform moles occur in about one in 1000 of the pregnancies of European women. These are masses of placental tissue gestated without a fetus, although mesenchyme and erythrocytes may develop within the mole. Women with molar pregnancies have a high incidence of circulating antibodies against placental HLA antigens; higher in fact than women with normal pregnancies. As the non-trophoblast content of the conceptus is less than in a normal pregnancy, this is surprising unless sensitization is caused by HLA antigens of the trophoblast (Beer and Billingham, 1976; Beer *et al.*, 1972).

Another piece of evidence which suggests that human trophoblast cells carry some antigens recognizable by lymphocytes—although not necessarily HLA antigens—is the cytotoxic response of lymphocytes to trophoblast cells with which they are cultured *in vitro*.

Maternal antibody against alloantigens characteristic of the father, whether evoked by fetal or by trophoblast cells, does frequently form during pregnancy, and can be detected in the circulation of the mother (Loke, 1978). None is found in the umbilical cord blood of the fetus. Fetal cord blood does contain antibodies other than those specific to paternal or fetal alloantigens. In fact, quite large amounts of immunoglobulin are ferried into the fetus during pregnancy, and are vital for the well-being of the infant after birth (p. 150). Only the potentially harmful antibodies against fetal antigens are conspicuous by their absence from cord blood, and must somehow be selectively filtered out by the placenta.

Unlike the spongiotrophoblast of the mouse placenta, trophoblast in human placentae lacks HLA antigen; specific trapping by simple antigen–antibody combination seems unlikely to explain the elimination of antifetal antibody. Placental cells other than trophoblast may play some part, since stromal fibroblasts and macrophages (for instance) do carry HLA antigen, but their contributions seem unlikely to be adequate (Johnson *et al.*, 1980).

A further possible mechanism of selective antibody trapping has emerged from investigations of the distribution of immunoglobulin receptor molecules within the human placenta. These are receptors for the Fc region of IgG (p. 9) and will therefore be unselective with respect to antigen-combining specificity of the immunoglobulin molecules with which they interact. IgGFc receptors are present on many of the cells of the placental syncytiotrophoblast, being particularly abundant in the first two trimesters of pregnancy. They are located also on macrophages within the stroma of the trophoblastic villi, and on endothelial cells linking the fetal blood vessels of the placenta (Figure 6.8)—see Wood *et al.* (1978*a*, 1978*b*).

Syncytiotrophoblast Ig receptors are probably involved primarily in the uptake of maternal IgG for transfer to the fetus (p. 153); they have a high affinity for single immunoglobulin molecules. As the Fc fragment of the IgG combines with the receptor, transfer of antibody will be indiscriminate, depending only on concentration within the maternal blood serum. If immunoglobulin specific for paternal type antigens has been formed, it will therefore be taken up by syncytiotrophoblast receptors and transmitted across the trophoblast as readily as any other. Why, therefore, do antifetal antibodies not appear in the fetus?

The answer may lie with the remaining IgG receptors, those of the stromal macrophages and blood vessel endothelia. Unlike the syncytio-

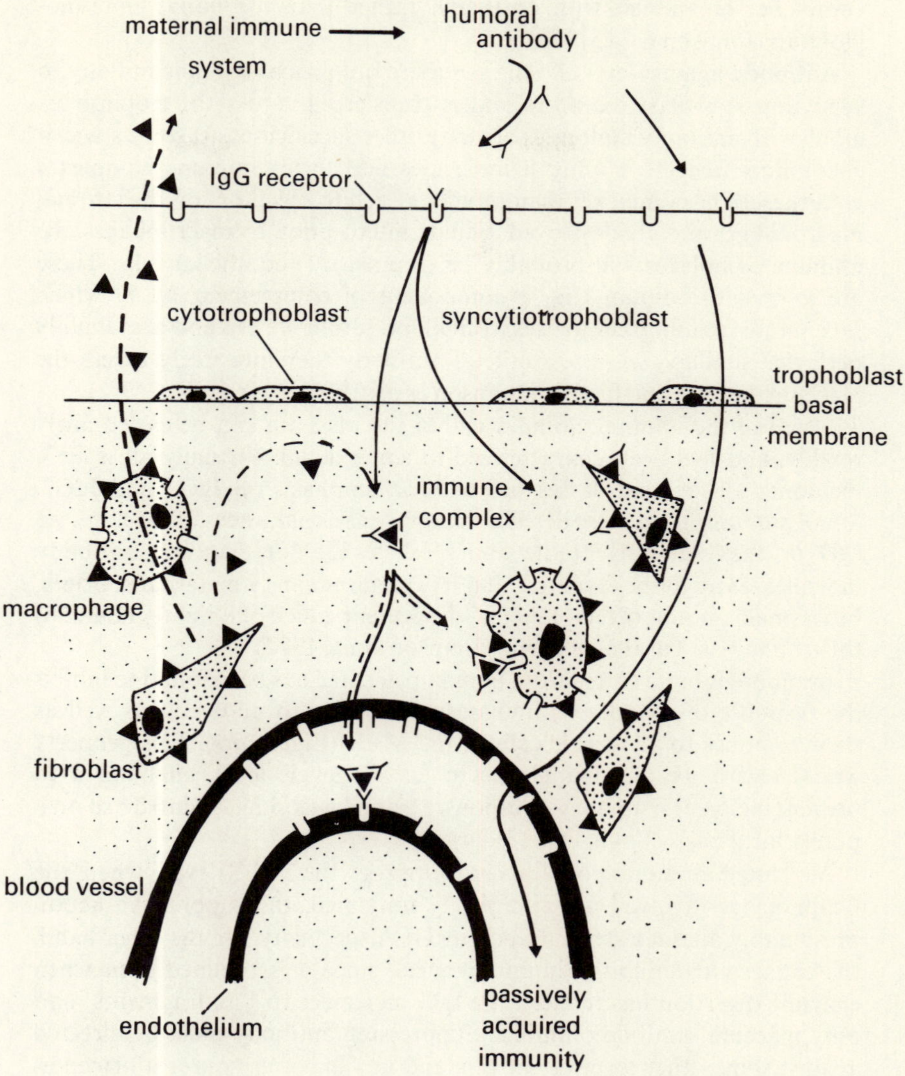

Figure 6.8 Possible mode of trapping of antifetal antibody by the human placenta. Antibody forms immune complexes with shed alloantigen and complement components; these are then prevented from entering the fetal circulation by Fc receptors on stromal macrophages and placental blood vessel endothelial cells (redrawn and adapted from Johnson *et al.*, 1980).

trophoblast receptors, these generally have a high affinity for aggregated IgG molecules. Experimentally, receptor affinity has been studied by using artificially aggregated IgG, but in normal circumstances, receptor affinity for aggregated immunoglobulin probably signifies affinity for immune complexes of antigen with antibody, rather than for native immunoglobulin (Johnson *et al.*, 1976).

Antibody against fetal alloantigen therefore probably binds initially to syncytiotrophoblast receptors and is transferred across the trophoblast along with antibody molecules of many other specificities. At some stage, it encounters free HLA antigen molecules and forms immune complexes with these. The complexes then rapidly attach to receptors on the stromal macrophages and are destroyed. Failing interception by macrophages, the immune complexes will probably be trapped by endothelial cells. These are known to contain Clq, a component of complement (p. 8) which precipitates immune complexes from blood serum *in vitro* and presumably performs similarly *in vivo*. Antifetal antibody therefore freely enters the placenta, but goes no further (Johnson *et al.*, 1980).

Much of the immunoglobulin within the placenta can be eluted fairly readily, and has been characterized in some detail. Virtually all is IgG, including all four subclasses but with an emphasis on IgG1 and IgG3. Small amounts of IgA are detectable, and occasional traces of IgM, but no IgD or IgE. Analysis of the allotypes (p. 135) of placental antibody indicates, as might be expected, that it is predominantly maternal in origin, but a small amount of fetal IgG may be present, much of it tightly bound to the trophoblast basal membrane (Johnson *et al.*, 1977).

Immunoglobulin eluted from human placentae has been found to inhibit the response of (unrelated) human lymphocytes to mitogens, as well as their response to allogeneic cells in the MLR (Figure 6.9). This property would clearly be a valuable one to an allogeneic fetus, enabling it to prevent incipient lymphocyte responses from developing, and to do so on a purely local basis (Faulk *et al.*, 1974).

As eluted immunoglobulin can suppress the MLR even when the lymphocytes involved are completely unrelated, the suppressive action presumably does not depend upon its HLA specificity. On the other hand, the capacity to inhibit is immunological, since it is retained even when enzymic digestion has reduced the IgG molecules to Fab fragments; and only placental antibody inhibits. Suppressing antibody must be directed against some other form of antigen, and in view of its source a placenta-specific antigen is indicated.

Human placental antigens have been investigated in some detail by Faulk and several collaborators, using antisera prepared in rabbits against a range of fractions of placental material. There appear to be two major antigens, or groups of antigens, involved. Faulk designates these TA_1 and TA_2 (TA = 'trophoblast antigen'). Their chemical nature is unknown, but TA_1 has been shown not to be identical with any of the placenta-associated glycoprotein antigens already described (Faulk *et al.*, 1978; Johnson *et al.*, 1980; McIntyre and Faulk, 1979*a*, 1979*b*).

Rabbit anti-serum TA_1 cross-reacts with certain proliferating human

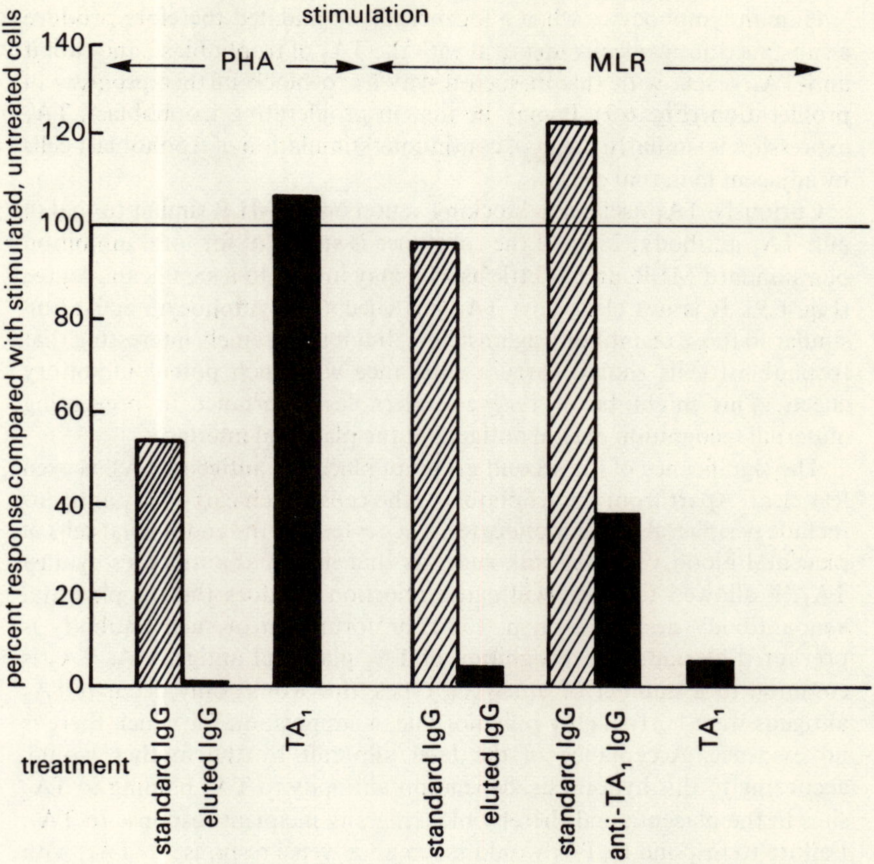

Figure 6.9 Effects of IgG eluted from human placentae (250 μg), rabbit antisera against TA_1 antigen (diluted 1:50), and TA_1 (25 μg) on human lymphocyte response to mitogen (PHA) and to allogeneic cells in the MLR. The response in terms of ^{3}H-thymidine incorporation, is here expressed as a percentage of the incorporation by stimulated lymphocytes not treated with IgG or TA_1. The effects of the treatment are compared as appropriate with those of a standard IgG preparation (based on data from Faulk *et al.*, 1974; McIntyre and Faulk, 1979*a*, 1979*b*).

cell lines cultured *in vitro*; and also with a proportion (3–5 %) of (unrelated) human lymphocytes shortly after these have been stimulated in the MLR. It does not cross-react with resting lymphocytes, nor with mitogen-activated lymphocytes; nor, indeed, with any other cell type. TA_1 expression therefore appears to reflect a cellular recognition mechanism involved in a lymphocyte response to allogeneic cells. To some extent its expression depends on the nature of the recognition. Lymphocytes of both rhesus monkeys and baboons interact vigorously with human lymphocytes in the MLR, both as stimulators and responders. Only with the rhesus monkey cells do human lymphocytes appear to produce TA_1.

Human lymphocytes when allogeneically stimulated therefore produce a substance antigenically identical with the TA_1 of trophoblast, and rabbit anti-TA_1 reacts with this in such a way as to block further progress of proliferation (Fig. 6.9). It may be that in proliferating trophoblast, TA_1 expression is similarly a sign of continuous stimulation of trophoblast cells by adjacent maternal cells.

Curiously, TA_1 itself has a blocking action on the MLR similar to that of anti-TA_1 antibody; 25 μg of the substance is sufficient for total inhibition of a standard MLR, and as little as 4 μg may inhibit to a significant degree (Fig. 6.9). It is not clear how TA_1 has effects on lymphocyte activation similar to those of antibody against TA_1, but it is extremely interesting that trophoblast cells should carry a substance with such potent inhibitory effects. This might be of very considerable importance in preventing maternal recognition of fetal antigens at the placental interface.

The significance of the second group of placental antigens, TA_2, is even less clear. Apart from the trophoblast, the cells which carry TA_2 antigens include peripheral blood monocytes, leucocytes and the endothelial cells of placental blood vessels. Faulk suggests that maternal antibodies against TA_1, if allowed to form, will cause abortion, as does the rat placental xenoantibody described on p. 138. The formation of such antibody is prevented by maternal recognition of TA_2 placental antigens. As TA_2 is common to a number of other cell types, this would only occur if TA_2 antigens were fairly highly polymorphic, a supposition for which there is no evidence. Acceptance of the fetal allograft by the mother would, according to this hypothesis, depend on antibody to TA_2 binding to TA_2 sites in the placenta and thereby blocking any incipient response to TA_1. Failure to respond to TA_2 would mean an adverse response to TA_1, with abortion as its probable sequel.

Many parts of this hypothesis rest, at present, on extremely speculative interpretations of rather slender evidence. It is, however, becoming evident that the placenta is the key to fetal survival, and that some remarkable interactions between mother and fetus probably take place there.

Specific receptors for transferrin, and transferrin itself, are also present on the surface of human placentae. As transferrin is apparently not synthesized by placental cells, it is presumably accumulated there through being trapped by surface receptor molecules. Transferrin is an iron-sequestering protein, probably involved in transfer of the considerable amounts of iron (up to 5 mg per day) demanded by the developing human fetus. Its iron-sequestering activity may also be usefully bacteriostatic, by depriving microorganisms of the element (Faulk and Galbraith, 1979; Johnson *et al.*, 1980).

Binding of maternal transferrin to the syncytiotrophoblast surface may also have protective effects against the maternal immune response. Accumulation of adsorbed protein of maternal origin will either conceal antigenic sites on the placental surface, or positively disguise the surface as maternal in nature. There is no evidence that such a mechanism is important, but as some helminth parasites are thought to survive by such a deception, it is conceivable that the fetal parasite does likewise.

Yet another intriguing possibility is suggested by the fact that stimulated lymphocytes require transferrin in order for cell division to proceed past a certain point in the cell cycle. By making transferrin unavailable, syncytiotrophoblast may therefore locally prevent lymphocytes from responding fully to antigen. There is also some reason to believe that iron-containing proteins such as transferrin may be important in regulating the distribution and migrations of lymphocytes. Lymphocyte traffic in the vicinity of the placenta might therefore be affected by the transferrin receptors of trophoblast.

Syncytiotrophoblast cells have also been shown, by immunofluorescence, to contain the protein transcortin. This is found also in the circulation, and rises markedly during pregnancy, due to increased synthesis in the liver under the influence of high oestrogen levels. Lymphocytes carry a specific cortisol receptor, which has been shown to be virtually identical with transcortin, and their content of this substance increases during pregnancy. Placental transcortin can block the response of maternal lymphocytes to mitogens such as PHA, and is even more potent in this respect than the plasma transcortin. Hence either directly, or indirectly by means of its affinity for (immunosuppressive) cortisol, placental transcortin may supplement other local immunosuppressive mechanisms (Werthamer *et al.*, 1976).

Some further ideas on the role of the placenta stem from the comparison of trophoblast with tumour cells. Many features are common to both; one of the most interesting is their survival in immunologically hostile environments. When malignant mouse tumour cells are cultured *in vitro* with macrophages, the macrophages are repelled. Trophoblast cells have a similar ability. The relevance of this mechanism to normal pregnancy has not been clarified, but the ability to hold maternal macrophages at bay might be a useful ancillary skill for trophoblast cells to have at their disposal (Fauve *et al.*, 1974).

Another feature shared by human trophoblast cells and some tumours is the presence of surface HCG. This might have an immunoprotective capacity distinct from the immunosuppressive role already discussed (p. 100). Treatment of human skin before grafting it to a guinea pig greatly prolongs survival of the graft (Borland *et al.*, 1975); placental HCG might perform in an analogous way. Coating of cells with a glycoprotein such as HCG, rich in sialic acid, is likely to cause a net increase in surface negative charge. Lymphocytes and trophoblast cells (and some tumours) are characterized by intrinsically high surface charge compared with other cells, and this is especially true of lymphocytes during human pregnancy

(p. 100). Surface HCG would therefore tend to minimize contact between lymphocytes and trophoblast cells by mutual electrostatic repulsion (McManus *et al.*, 1976; Whyte, 1978). The enigmatic finding that maternal lymphocytes—but not, apparently, unrelated lymphocytes—stimulate HCG production by trophoblast *in vitro* may suggest some form of defensive response by trophoblast to immunological hazards. How trophoblast might recognize lymphocytes and discriminate between maternal and unrelated ones, is completely unknown (Dickman and Cauchi, 1978). However, whatever the mechanisms involved, trophoblast cells do seem to have the potential for local regulation of some aspects of the immune system (Barg *et al.*, 1975; Pavia *et al.*, 1981).

There are therefore many different forms of possible maternofetal interaction within the placenta. Some are of established importance; with others, the significance is highly speculative. It seems highly probable, however, that the mammalian placenta, and the transactions that take place within it, are crucial for survival of the antigenically alien fetus.

6.7 Harmful immune reactions against the fetus

Deleterious immune responses against the fetus are in practice rare. One exception is the form of haemolytic disease of the newborn known as Rhesus disease; until quite recently, this accounted for perhaps half of perinatal mortality.

Rhesus haemolytic disease is due to the formation by a mother of antibody against red blood cell (erythrocyte) antigens of her fetus, the passage of this antibody across the placenta into the fetus, and the consequent destruction of fetal erythrocytes. Since the mortality generally occurs around the time of birth, antibody transfer evidently takes place at the end of pregnancy.

As a mother can only produce antibody to antigens which she does not herself possess, only Rhesus positive fetuses in Rhesus negative mothers are at risk. About 85% of the European population is Rh-positive, so a high incidence of Rhesus disease would be expected. In fact, the incidence is a great deal lower than expected, for a number of reasons.

Rh antigens, unlike those of the ABO blood group system, are found only on erythrocytes. Although their chemical nature is unknown, they are known to be the products of a genetic complex which apparently consists of three loci, each of which has two principal alternative alleles. These are termed C/c, D/d and E/e. Of the three, only D is significantly immunogenic. C and E may, however, modify its expression: for example, CDE is a less immunogenic combination than cDE. It has been suggested that the presence of C antigen reduces the number of D determinants at the erythrocyte surface (Race and Sanger, 1975).

Passage of fetal erythrocytes into the maternal circulation is a common event in human pregnancy (p. 112) and, because fetal cells contain a form of haemoglobin not found in adult cells, is relatively easy to monitor. Passage may start as early as the 8th week of pregnancy; during the final three months the incidence is 30–40%, rising to 50% after birth (or even higher, depending on the obstetric measures employed). Judging from the number of fetal erythrocytes found, the amounts of blood transferred are small; 73.5% of mothers in one survey had acquired the equivalent of 0.05 ml or less, 24.6% between 0.05 and 5 ml, and less than 2% more than 5 ml. However, since as little as 0.5 ml is an effective sensitizing dose, a significant proportion of mothers are likely to become sensitized to Rh antigen.

A peculiarity of the Rh system is that a first Rh-incompatible fetus only rarely evokes antibody production during pregnancy, despite the entry of fetal blood into the mother. This is because the antibody response to Rh antigens is extremely sluggish. When male Rh-negative volunteers are injected with D antigen, antibody appears only after weeks or even months. In the analogous situation of pregnancy, antibody therefore appears only some time after birth, when the baby is no longer vulnerable. A second Rh-positive fetus will, however, be exposed either to preexisting antibody or to antibody produced in a more rapid secondary response.

Another factor in the fetus's favour is the ABO blood group system. Where A antigens are present on the erythrocytes, anti-B antibody is naturally present in the blood serum, and *vice versa.* Thus when an erythrocyte carrying both Rh and group A antigen enters the circulation of a Rhesus negative mother of group B or group O, the anti-A antibody of the mother has prior claim, destroys the erythrocytes, and deprives the Rh antigen of the opportunity to induce anti-D antibody. ABO incompatibility thus gives some protection against Rh incompatibility, virtually halving the likelihood of sensitization to Rh antigen.

Although some anti-D IgM may be produced, particularly in the early stages of sensitization, the major component of anti-D is IgG: this is unfortunate, since IgG crosses the human placenta whereas IgM does not (p. 154). Most of the natural antibodies of the ABO system are IgM, which accounts for the rarity of fetal haemolytic disease based on ABO sensitization. Rhesus antibody is not complement-fixing, possibly because of the sparseness of D antigenic sites on the erythrocytes (about 1.2×10^4, compared with, for instance, some 5×10^5 MHC determinants on a lymphocyte). How fetal erythrocytes are destroyed is therefore not altogether clear; probably antibody attachment marks them out for subsequent disposal by macrophages. IgG1 and IgG3 constitute a large production of the anti-D antibody, and are known to be effective in this respect.

Although it is rare for anti-D antibody to be produced during a first incompatible pregnancy, this does occasionally occur. In such unexpected cases, the maternal grandmother is frequently Rh positive. It seems possible therefore that Rh sensitization can occur in the reverse direction: if a Rhesus positive woman bears a Rh-negative child, maternal erythrocytes may occasionally enter the fetal circulation and sensitize the fetus. When the baby, if female, grows up and conceives a Rh-positive fetus, she may therefore develop a secondary response. In this case her own first baby may be affected. Alternatively, it has been suggested that some of the cases where sensitization is predictable but fails to take place may be due to a similar cause. If a Rh-negative fetus can become sensitized to maternal Rh antigen, she might also perhaps become permanently tolerant of the antigen and fail to respond when pregnant with an incompatible fetus.

The elucidation of the causes of Rhesus haemolytic disease has been followed by the development on clinical methods for its treatment. Methods for detecting anti-D antibody in maternal blood, and analysis of amniotic fluid for haemoglobin breakdown products, have made it possible accurately to identify fetuses at greatest risk before any damage becomes serious. Replacement of the baby's blood by massive transfusions after birth (or even before) can then be carried out.

A more elegant procedure is the treatment of the mother with anti-D antibody antiserum. The formation of anti-D antibody is exactly what must be prevented, so this procedure may, at first sight, seem perverse. The effect of administering anti-D from an external source, however, is to block the prevention of endogenous anti-D, in much the same way as the natural occurrence of antibodies of the ABO system does in appropriate cases. Anti-D is usually given just after the birth of the first Rh-positive baby and, by eliminating Rh-positive erythrocytes and preventing their stimulation of the antibody-producing system, protects a second incompatible pregnancy. The artificially administered antibody will itself be broken down and eliminated long before the second conception could take place.

The cumulative effect of these, and other measures, has been remarkable: from being a major cause of infant mortality, Rhesus disease has become of minor, and dwindling, importance (Beer and Billingham, 1976; Loke, 1978; Tovey and Maroni, 1976).

Maternal immune reactions against fetal histocompatibility antigens have been implicated in the condition known as pre-eclampsia. Diagnosis of this disorder of pregnancy is somewhat subjective, and almost certainly a number of different causes contribute to the development of a constellation of symptoms. Differences in definition make it virtually impossible to reconcile the many attempts to clarify the situation (Jenkins, 1976; Loke, 1978; Marti and Herrmann, 1977; Redman, 1980).

The clinical features of pre-eclampsia include a rise in blood pressure, associated with defective kidney function, activation of the blood clotting system, and thrombosis. The condition can prove fatal, with the main cause of death cerebral haemorrhage; in less extreme forms—affecting perhaps 5% of pregnancies—the major effects may be on the fetus: thrombosis in small blood vessels can impede the placental blood circulation and lead to serious fetal distress.

Numerous studies have implicated immune factors in pre-eclampsia, and about as many have rejected it. Evidence in favour is that the incidence is about ten times higher in first pregnancies than in subsequent pregnancies by the same father. This is interpreted as an indication that exposure to fetal antigens during the first pregnancy gives protection against pre-eclampsia in subsequent ones. Supporting this is the finding that the protective effect does not apply to second pregnancies with changed partners, and exposure to a different set of histocompatibility antigens. As would be predicted from the hypothesis of an immunological basis for pre-eclampsia, affected women are according to some surveys likely to be more HLA-incompatible with the fathers of their children than unaffected mothers. Other surveys come to the opposite conclusion.

Finally, if HLA alloantigen exposure during a first pregnancy reduces the risk of pre-eclampsia in subsequent pregnancies, it follows that the first pregnancy would itself be protected by exposure to appropriate HLA antigens before conception. HLA antigens are carried on spermatozoa, and there is evidence of sensitization to HLA antigens following coital exposure (p. 66). Although alternative interpretations are possible, there is statistical evidence that the amount of unprotected sexual intercourse with the father before conception is related to the probability of pre-eclampsia developing during a first pregnancy; the greater the coital exposure to spermatozoa, the less the chance of the pregnancy being affected by pre-eclampsia.

Although all of the evidence is circumstantial, and much of it contradictory, there is therefore some reason to believe that maternal responses to antigens of the fetus can at least contribute to serious complications of pregnancy. Indeed, it would be surprising if there was no involvement of immune responses in diseases of pregnancy; the remarkable rarity of maternal attacks on the fetus is testimony to the efficiency of the protective mechanisms which have evolved with the evolution of viviparity in mammals.

CHAPTER SEVEN

PASSIVE TRANSFER OF IMMUNITY

7.1 Routes of transfer

Although mammalian fetuses are generally capable of antibody synthesis well before birth, little antibody is in fact produced. This is doubtless due partly to depression of the immune response by various immunosuppressive measures (p. 134). However, it stems also from the fact that the fetus is protected from exposure to foreign antigens (other than those of the mother herself). After birth, this is no longer the case, and the newborn mammal is suddenly and simultaneously exposed to a wide spectrum of microorganisms, many of them pathogenic. An immunologically naïve animal, possibly with some residual depression of its capacity to respond, would be extremely vulnerable to this onslaught.

The relative lack of endogenous antibody is compensated for by the acquisition of a range of maternal antibodies directed, presumably, against currently prevalent pathogens. This passively acquired immunity is of great value in tiding the neonate over its most vulnerable first few days of life in the outside world. Thereafter the passively acquired maternal antibody progressively disappears from the circulation, to be replaced by antibody actively produced by the young animal itself.

In some species, antibody is acquired while the fetus is still in the uterus; in others it is secreted by the mammary glands and acquired by suckling. Table 7.1 lists the major means of antibody transfer in a number of species (Brambell, 1970).

7.2 Prenatal transmission of immunity

In the rabbit, although some antibody may pass into the fetus earlier, the bulk of the transfer takes place during the second half of the gestation

Table 7.1 Routes of transfer of passive immunity from mother to young (data from Brambell, 1970)

Species	*Prenatal*	*Route*	*Postnatal*	*Period*
Horse	0	–	+++	24 h
Pig	0	–	+++	24–36 h
Ox, goat, sheep	0	–	+++	24 h
Dog, cat	+	?	++	1–2 d
Mouse	+	y.s.	++	16 d
Rat	+	y.s.	++	20 d
Guinea pig	+++	y.s.	0	–
Rabbit	+++	y.s., p	0	–
Human	+++	p	+	24 h

y.s. = yolk sac, p = placenta.

Of the species listed, horses, pigs, oxen, goats and sheep have epitheliochorial forms of placenta (p. 88), with a notional six intervening tissue layers (although the maternal epithelium may make a negligible contribution); dogs and cats have 'endotheliochorial' placentae in which maternal connective tissue and epithelia are absent; the remaining species have forms of haemochorial placentae. The rabbit placenta is initially epitheliochorial and loses layers progressively, while in the placentae of rabbits and of the rodents listed further erosion of trophoblast and fetal connective tissue may occur.

In the case of the human neonate, much antibody enters the gut but only a small amount is taken up from there into the circulation.

period, which is of 30 or so days. Rates of transmission rise from 20 days after mating to a peak at 26 days, and subsequently decline. The speed of transmission is shown by the appearance in the fetus of labelled (rabbit) antibody within an hour of its injection into the pregnant female, and equalization of fetal serum concentrations with those of the mother within 4 hours. The immunoglobulin classes transmitted include IgM as well as IgG; this contrasts with the majority of mammal species, where IgM generally fails to pass into the fetus (p. 113).

Although the chorioallantoic placenta might appear an obvious route of transmission, in fact it is of, at most, minor significance; only recently has any antibody transmission by this route been firmly established. The main means of transfer is the yolk sac which is, in rabbits, a relatively large structure, in direct contact with the uterine lumen for much of pregnancy. Around the 16th day of pregnancy, the outermost layer of the yolk sac disappears, exposing the inner endoderm of the yolk sac directly to the contents of the lumen. Thereafter, antibody uptake is rapid. Surgically preventing passage of blood between the yolk sac endoderm and the fetus also prevents the uptake of antibody from the lumen or from the maternal blood system; clearly the yolk sac is the major source of maternal antibody for the fetus.

Not all proteins in the maternal serum pass into the fetus with equal ease. Rabbit immunoglobulin is favoured over that of other species, and intact immunoglobulin molecules over fragments (Figure 7.1). How do proteins enter the fetus, and how is the specificity of the transport

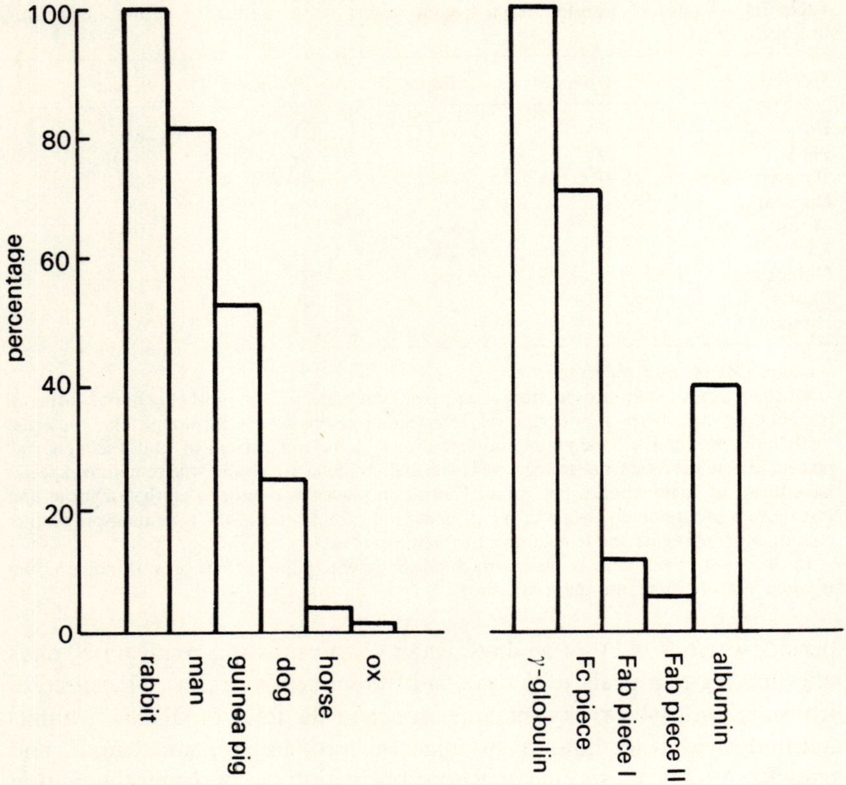

Figure 7.1 Selective transmission of immunoglobulin across the yolk sac of the rabbit embryo. Left—immunoglobulin prepared in the species named; right—transmission of entire immunoglobulin molecules, of various fragments, and of albumin (redrawn from Brambell, 1970).

achieved? Large pinocytotic vesicles have been observed to form at the luminal surfaces of yolk sac cells. These, however, would probably be unselective in their uptake. To explain the evident selectivity of transport, Brambell proposed that rabbit antibody attaches to cell surface receptors and is, in a bound state, incorporated into the membranes of pinocytotic vesicles as they form. Other constituents of the uterine fluid would also be incorporated, but not bound to membrane receptors. After their formation, the vesicles are thought to merge with lysosomes. This would result in the vesicle contents being catabolized; the rabbit immunoglobulin, it was postulated, was protected by being bound to its specific receptors and could then be disgorged on the fetal side of the cell (Brambell, 1970).

A more recent view is that all of the contents of the large pinocytotic vesicles are catabolized, including any rabbit immunoglobulin present, and that the transport of intact immunoglobulin is the province of a different and specialized form of vesicle. These appear first at the surface of the yolk sac cells as 'coated pits', the coating being a fuzzy superficial layer largely made up of a protein known as clathrin. Receptors for the Fc region of rabbit immunoglobulin—specifically, for the C_{H^2} domain (Johanson *et al.*, 1980) also accumulate in the pits and cause immunoglobulin to bind here. As the pits are internalized, they become 'coated vesicles', distinguishable from the larger, and uncoated, vesicles which merge with lysosomes. Coated vesicles pass through the cell intact and discharge their immunoglobulin intact on the fetal side of the membrane (Figure 7.2). The antibody transferred is mainly IgG, but some IgA transfer also takes place (Goldstein *et al.*, 1979; Hemmings, 1976, 1979).

In the mouse, rat and guinea pig, the prenatal acquisition of immunity is relatively less important; however, the mechanism involved is similar, with the yolk sac as the main route of transmission and the placenta playing an ancillary role.

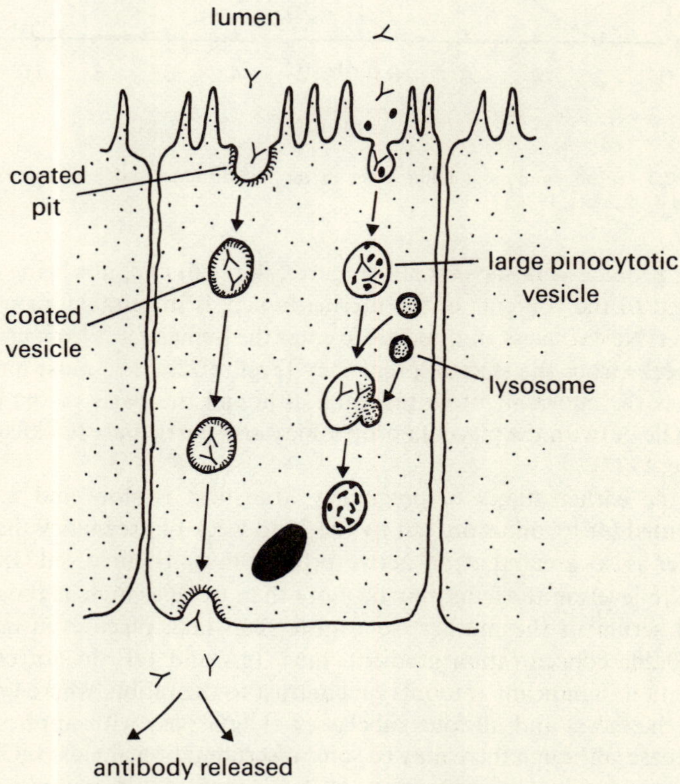

Figure 7.2 Possible mode of transmission of immunoglobulin across the yolk sac membrane of the rabbit (based on Goldstein *et al.*, 1979; Hemmings, 1976, 1979).

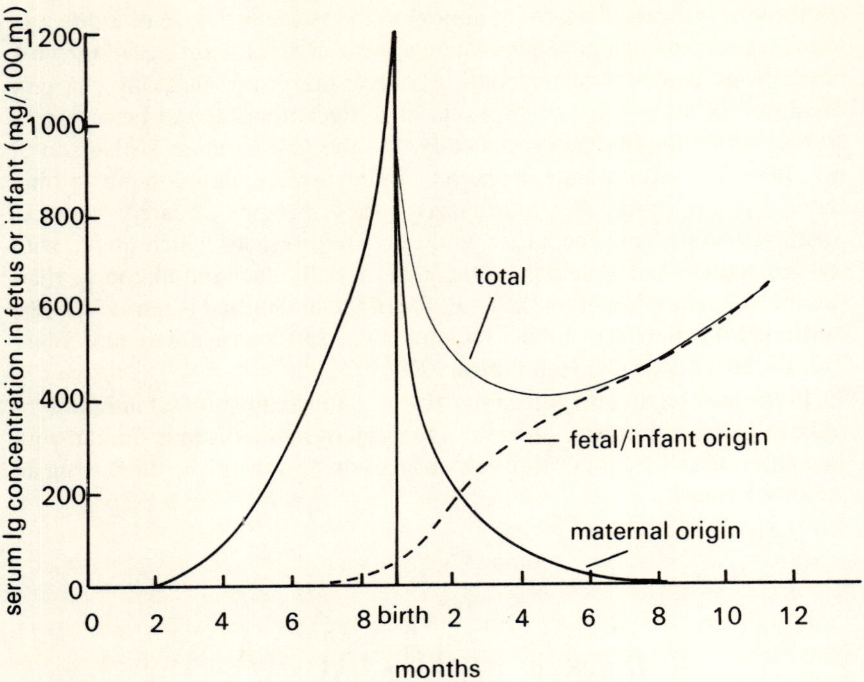

Figure 7.3 Serum immunoglobulin levels in the human fetus and infant (redrawn and adapted from Jones, 1976).

The primate yolk sac is small (Figure 5.4, p. 86) and, not being directly exposed to the contents of the uterine lumen, is unsuitable for antibody transfer. Nevertheless, antibodies do enter the human fetus from as early as 7–8 weeks from the start of pregnancy (Figure 7.3). The most important route is the chorioallantoic placenta, although the parts of the chorion which lie outwith the placenta proper may also participate (Ockleford and Whyte, 1977).

In the earlier stages of pregnancy, transport is slow and might be accounted for by diffusion, but by the 22nd week of pregnancy the rate of transfer is so great that an active process must be involved. Immunoglobulin levels in the fetus may be more than twice as high as those in the blood serum of the mother, so transfer can take place even *against* a formidable concentration gradient. IgM, IgA and IgE do not cross the placenta in significant amounts (in contrast to the rabbit, where IgM does enter the fetus) and all four subclasses of IgG pass with approximately equal ease, although there may be some discrimination against IgG2 which constitutes a lower proportion of fetal than adult immunoglobulin (Pitcher-Wilmott *et al.*, 1980).

Receptors for the Fc region of IgG have been identified on syncytiotrophoblast cells of the placental villi, as well as on other cells within the placenta (p. 140). 'Coated vesicles' like those of the rabbit or mouse yolk sacs have also been found within human placentae. It seems most likely that the mechanism of transfer across the human placenta is closely similar to that across the rabbit yolk sac endoderm (Wood *et al.*, 1978*a*).

7.3 Passive transfer of immunity after birth

In many species of mammal, the passive transfer of immunity continues after birth, with antibody being supplied to the infant in the milk. In some species, this is the only means of acquisition of antibodies from the mother. For the first few days after birth, the mammary glands secrete a specialized form of milk called colostrum; compared with milk proper, colostrum is rich in immunoglobulin (Table 7.2a) and in many species forms the dominant, if not the only postnatal source of antibodies. In other species, antibody transfer in the definitive milk is significant, and may be so for the greater part of the period of lactation (Richards, 1979; Watson, 1980).

Table 7.2a illustrates the antibody composition of cow colostrum, compared with that of milk and with blood serum. The dominant immunoglobulin is IgG, particularly IgG1; the concentration in colostrum may be as much as five times greater than in maternal blood serum. This indicates that it is actively transported across the secretory epithelium of the mammary gland. The process is apparently similar to that already described for antibody passage across yolk sac and placental tissues.

The gut of a calf secretes acids which might denature immunoglobulin molecules, as well as an assortment of proteolytic enzymes. Several factors militate against destruction of antibody molecules before they can be absorbed. Firstly, colostrum has considerable buffering power against extremes of pH. Secondly, IgG1—the predominant immunoglobulin in

Table 7.2 Immunoglobulin content (in mg ml^{-1}) of (a) bovine blood serum, colostrum and milk; and (b) the colostrum of a number of species of mammal (data from Watson, 1980)

(a)	*IgG1*	*IgG2*	*IgA*	*IgM*
Blood serum	14.0	13.0	0.4	3.8
Colostral whey	40.0–80.0	2.5	4.5	6.0
Milk whey	0.4	0.06	0.1	0.1

(b) *Species*	*IgG*	*IgA*	*IgM*
Sheep	60.0	2.0	4.1
Cow	50–90.0	4.5	6.0
Pig	57.0	10.0	2.7
Rabbit	2.4	4.5	0.1
Human	0.2	18.0	0.8

cow colostrum—appears to be intrinsically more resistant than other immunoglobulin types to proteolytic digestion by the gut enzyme chymotrypsin; and, finally, a trypsin inhibitor is present in colostrum which will inhibit trypsin-mediated proteolysis.

Uptake from the gut of the suckling calf is apparently unselective, but by virtue of the selective nature of secretion, and of immunoglobulin survival after ingestion, the young animal acquires much more IgG1 than it does antibodies of other classes and subclasses. Absorption ceases with 'gut closure' about 36 hours after birth, and the replacement of the absorptive cells of the intestine by cells which are unable to absorb intact antibody molecules.

To the calf, the importance of this process is considerable, and survival of calves which have been deprived of colostrum is unlikely. Protection is obtained against a variety of viral and bacterial pathogens, of which one of the most dangerous is the gut bacterium *Escherichia coli.* This suggests that part at least of the beneficial action of the transferred immunoglobulin may be within the gut, before transfer through the wall of the intestine into the circulation (Watson, 1980).

In other species, the dominant colostral antibody is secretory IgA (Table 7.2b) although some IgG is also present. In the rat, antibody is taken up in the proximal small intestine, probably by a selective mechanism like that already described, with specific surface immunoglobulin receptors and coated vesicles. Transfer ceases at around 20 days of age, at a time when antibody is still present in the milk. Adrenal corticosteroids appear to play some part in the process of 'gut closure' (Rodewald, 1973).

Humans, like rats, have secretory IgA as the dominant antibody of colostrum and milk. Serum antibody levels in the human infant fall sharply after birth (Figure 7.3), and the rise which then follows is the result of synthesis by the infant, not transfer from the mother. Although the infant does in fact ingest quite large amounts of maternal antibody, very little of it passes into the blood. In one study, breastfed babies were given colostrum containing IgA against poliovirus, at various times after birth. Three—all given the IgA within 24 hours of birth—showed transfer of the antibody into their blood serum; this amounted to only 10–12% of the total received. Approximately 60% was voided in the faeces, and the remainder unaccounted for. Presumably part of this was retained on the mucosal surface of the gut, and part underwent proteolysis (Ogra *et al.*, 1977).

Most of the passively acquired IgA must therefore function within the gut of the newborn. The effects of the transferred antibody, in some instances greatly potentiated by other colostral components such as lactotransferrin (cf. transferrin, p. 144) are largely on potentially harmful gut microorganisms such as *E. coli* (Stephens *et al.*, 1980). Comparison of the mortality rates of breastfed and bottlefed infants emphasizes the importance of this protection: the former have mortality rates less than one-fifth those of the latter (10.2 compared with 57.3 deaths per 1000 births, in one U.K. survey)—see Loke (1978).

Human colostrum (like that of other species) contains considerable numbers of cells, ranging from 10^5 to 10^7 per ml, with great variation both within and between mothers. Many of these cells are polymorphonuclear leucocytes, the major phagocytic cells of the blood (40–60 %); of the remainder, 30–40 % are macrophages, and up to about 10 % lymphocytes with T and B cells in roughly equal proportions. Many of these cells presumably protect the mammary gland itself, as well as conferring benefits on the infant. Both polymorphs and macrophages are actively phagocytic, and the latter may in addition have a cytostatic effect on certain forms of tumour cell (Balkwill and Hogg, 1979; Beer and Billingham, 1976; Crago *et al.*, 1979).

The behaviour of colostral lymphocytes is not identical to that of circulating blood lymphocytes. B cells committed to IgA, IgG and IgM synthesis are all present. When milk lymphocytes are tested *in vitro*, antibody synthesis can be stimulated only in the IgA-producing cells. Peripheral blood lymphocytes in similar circumstances respond with all three classes of antibody. When blood lymphocytes in culture are treated with the supernatant of colostrum they, too, respond only with IgA production. This suggests that colostrum contains a soluble factor which regulates immunoglobulin synthesis, either restraining synthesis of IgG and IgM, or stimulating that of IgA. The factor appears to be present in milk only shortly after the start of lactation (Pittard and Bill, 1979).

The lymphocytes which infiltrate the epithelium of the mammary glands during pregnancy and lactation, and which then pass into the secreted colostrum and milk, are not a random sample of the lymphocyte population of the mother. A high proportion of them are committed to IgA production. Milk lymphocytes are relatively unresponsive to mitogens and to stimulation by allogeneic cells, unlike those found in the circulation. Instead, they demonstrate a high degree of responsiveness to antigens characteristic of gut pathogens. An example of this is the K1 capsular antigen of *Escherichia coli.* Milk cells respond to this antigen vigorously, with the production of IgA; peripheral blood lymphocytes fail to respond. *E. coli* K1 is found in the gut contents of some 44 % of post-partum women, and is highly pathogenic in newborn infants. About 84 % of infant meningitis attributable to *E. coli* is caused by this strain. This evidence suggests that milk B cells are derived preferentially from gut-associated lymphocytes of the mother, and that the antibodies they produce are those most likely to be protective against pathogens in the gut of the newborn infant (Parmely *et al.*, 1976; Seelig, 1980). Transfer of complement components may also be important (Renshaw and Gilmore, 1980).

It has been suggested that maternal lymphocytes can pass intact through the infant's intestine wall and permanently colonize its lymphoid organs. The evidence is based largely on the apparent lactational transfer of immunity of types known to be cell-mediated. The acquisition by human infants of tuberculin sensitivity is one example. The duration of such passively acquired cellular immunity appears to be brief: in one study, sensitivity was demonstrable in peripheral blood lymphocytes collected from breastfed infants at 5 weeks of age, but no longer so at 12 weeks, although breastfeeding was continuing at that time (Slobodian *et al.*, 1980).

Another example of the possible transfer of cell-mediated immunity is the apparent acquisition of tolerance to allografts of Fischer strain skin by young rats of the Lewis strain fostered at birth on Fischer strain females. One result of this tolerance is that some 30% of the fostered rats die as a result of a 'wasting syndrome', probably as a result of attack by infiltrated Fischer strain lymphocytes on their tolerant host. Direct investigation has failed, however, to provide any clear evidence of the successful entry of maternal lymphocytes through the intestine wall of infant mammals; it is possible, therefore, that what is transferred is not whole cells but some form of soluble factor which affects the behaviour of the infant rat's own lymphocytes. Whatever the explanation, it is evident that complex immunological interactions between mother and young do not stop at the moment of birth (Beer and Billingham, 1976).

CHAPTER EIGHT

IMMUNOLOGICAL CONTRACEPTION

8.1 Strategies

The processes of reproduction normally proceed satisfactorily despite the immune system. Some forms of human infertility, nevertheless, do result from immune responses; these are discussed in sections 3.5 and 4.4. Male infertility can be caused by autoimmunity against spermatozoal or other seminal antigens, and female infertility by a woman's response to isoantigenic or alloantigenic components of semen. There have been reports also of female autoimmunity to steroidogenic tissues and to ovarian antigens, including antigens of both the zona pellucida and the cytoplasm of the oocytes; this, too, appears to be associated with infertility.

The quest for forms of contraception additional and alternative to those currently in use has motivated many investigations of the feasibility of duplicating, at will, situations like these: if spontaneous immune responses can impede fertility, can exploitation of immune responses to antigens of the reproductive system provide a means of deliberately controlling fertility?

An ideal form of contraception should fulfil four principal criteria. Firstly, it should be effective: 100% reliability may not be essential in the perspective of limitation of population growth, but to individual couples it is clearly crucial. Secondly, the method must be acceptable and convenient to use. Thirdly, it should have no significant side-effects; and, finally, it should be fully reversible.

The obvious attraction of immunological methods of contraception lies chiefly with the third criterion. The specificity of the immune system should make it possible, in principle, to achieve a precision lacking from, for instance, the artificial manipulation of steroid hormone levels. Judicious selection of a target antigen peculiar to the immune system should make it possible to suppress fertility without affecting other physiological systems. Obvious candidates for attention are the gamete-specific antigens of eggs or sperm, and certain of the reproductive hormones. In this chapter I discuss some recent progress, along these lines, in the development of immunological techniques of contraception.

8.2 Immunization against gamete-specific antigens

Immunization of male guinea pigs against testis homogenate or purified spermatozoal antigens can result in an autoimmune response affecting the testes, and the complete cessation of spermatogenesis (p. 43). The same is true of other species of mammal. Can this technique be extended to man?

An obvious problem is the availability of volunteer subjects for such experiments. The only subjects on whom immunization attempts have been carried out were men suffering from carcinoma of the prostate. In one group of volunteers, in whom castration was indicated as a necessary therapeutic measure, a single testis was removed from one patient and used as a source of antigen to inject into all members of the experimental group. Antigen, in the form either of testis homogenate or of an extract, was given

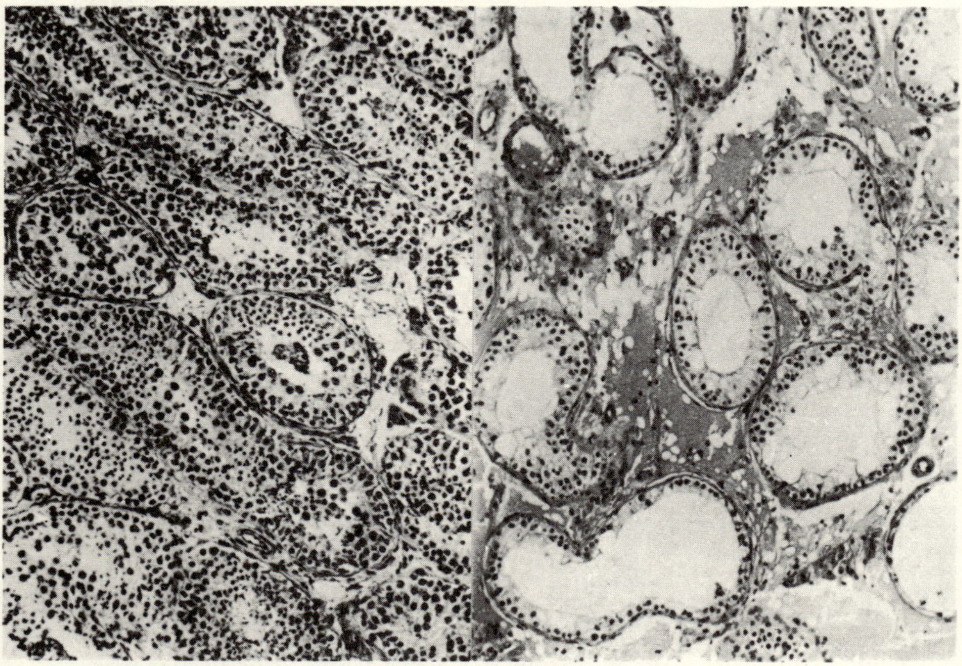

Figure 8.1 The effects on the human testis of immunization against sperm. Left—normal seminiferous tubules in transverse section; right—a similar view of a testis after immunization; germ cells are virtually absent from the tubules (from Mancini, 1971).

as a single injection in adjuvant. Over the following weeks, various signs of an immune response were detectable, including sperm-immobilizing antibody in the blood serum (p. 67). Examination of testis material following biopsy or castration showed varying degrees of disorder within the seminiferous tubules. Vacuoles appeared in the cytoplasm of Sertoli cells, with sloughing of germ cells from the seminiferous epithelium; this particularly affected spermatocytes, spermatids and spermatozoa, which virtually vanished from between 50% and 80% of the tubules (Figure 8.1). It seems inevitable that fertility would have been severely impaired, had this been put to the test (Mancini, 1971; Mancini *et al.*, 1965).

An acceptable form of immunization against spermatozoal antigens in men would probably involve the use of purified antigen preparations rather than simple homogenates or crude extracts. In either case, a source of antigen is required, and it can probably be assumed that the supply of voluntary human testis donors would be less than overwhelming. Antigens extracted from ejaculated spermatozoa might be available in sufficient amounts, or (if the spermatozoal antigens required were not species-specific) antigenic material from other species.

An alternative technique, which has been applied successfully to a range of species other than man, is to employ endogenous antigen. Talwar, working in India, has developed a method of injecting adjuvant material—Bacille Calmette Guérin (BCG), used as a vaccine against tuberculosis—directly into the testis, so breaching the blood-testis barrier (p. 34) and provoking a local autoimmune response. This has been done in rats, guinea pigs, dogs, rams and rhesus monkeys. Some results are shown in Figure 8.2; within two months of a single intratesticular injection of BCG, spermatozoa had virtually disappeared from the semen, and the few sperm remaining were immobile and of poor viability. Examination of the testes showed large-scale infiltration of the seminiferous tubules by leucocytes. Side-effects in most species—but not in the rhesus monkeys—included some pain, and swelling of the testes was common; on the positive side, Leydig tissue, blood testosterone levels and sexual behaviour were unaffected by the treatment. In several subjects, spermatozoa later reappeared in the semen, so the induced infertility appears, partly at least, to be reversible. There seems no good reason to suppose that this method would not work in men, although the direct testis injection might seem an unappealing proposition (Talwar, 1980; Talwar *et al.*, 1979).

The most promising of the purified antigen preparations so far tried on experimental animals is LDH-X, the sperm-specific form of the enzyme lactic dehydrogenase (p. 36). Immunization of female rabbits with mouse LDH-X in adjuvant leads to antibody production and a virtual halving of the conception rate, together with considerable post-implantation embryo mortality (p. 68). The lack of species-specificity of LDH-X suggests that if the technique was extended to humans the supply of antigen would

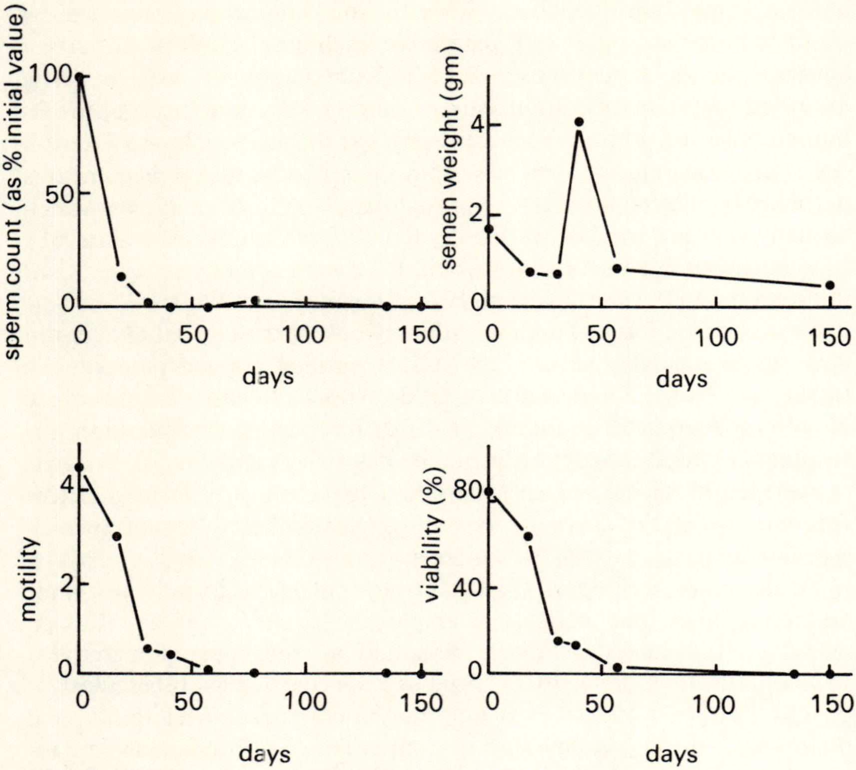

Figure 8.2 Effect of intratesticular injection of BCG on sperm count, motility and viability, and on weight of semen, in a rhesus monkey. The animal was injected a few days after the initial measurements (redrawn from Talwar *et al.*, 1979).

present no problems. Significant post-implantation mortality of embryos, on the other hand, probably would prove unacceptable; however, as LDH-X immunization in baboons reduces fertility without having this effect, humans might be similarly exempt (Goldberg, 1974; Goldberg and Lerum, 1972; Goldberg *et al.*, 1981; Kille and Goldberg, 1979).

Antigens peculiar to the ovary have also been described, the most significant being associated with the zona pellucida (p. 81)—see Sacco (1977). An advantage of the zona of a mature oocyte as a target for immunological attack would be that only a single cell, once every 28 days or so, need be considered; a very different prospect from attempting the immunological suppression of countless millions of spermatozoa.

For fertilization to occur, a sperm must first attach to the outer surface of the zona pellucida at a specific receptor site. Initial attachment is followed by penetration of the zona with the aid of the acrosomal protease acrosin. When anti-zona antibody is added to mouse, rat or hamster eggs *in vitro*, neither attachment nor penetration is likely to occur. Attachment of spermatozoa is probably prevented by the cross-linking of surface glycoprotein chains by the antibody molecules. This presumably masks the sperm attachment sites; a precipitate of anti-zona antibody is visible on the outer surface of affected zonae. Protease digestion of the zona material is also slowed by the cross-linking, so any spermatozoa which do attach will have difficulty in penetrating the zona; and even if fertilization is achieved despite these obstacles, the blastocyst may later fail to shed its zona. This would, *in vivo*, reduce the chances of successful implantation.

Anti-zona antibody, whether passively administered or actively induced, also reduces fertility dramatically *in vivo*. Cross-reactions between species are widespread so that, for instance, immunization with mouse oocytes can reduce the fertility of female rats. In one series of trials, 96% (44/46) of matings by actively immunized female rats failed to result in conception, with only one out of eight matings in a control group being unsuccessful. Infertility has been shown to last for over six months in a number of cases. Pig and human zonae similarly cross-react antigenically, so that plentiful antigen might be available should anti-zona immunization come to be used as a contraceptive technique in women (Aitken and Richardson, 1980; Boettcher, 1977; Talwar, 1980).

8.3 Immunization against reproductive hormones

The hormones which integrate the processes of reproduction are also obvious candidates for immunological intervention. The endocrine control of male and female reproduction is outlined in sections 3.1 and 4.1. In both sexes, the gonads produce steroid hormones which affect secondary sexual characteristics and sexual behaviour as well as gametogenesis. In both sexes, the pituitary gonadotrophins, glycoprotein in nature, control the gonads. Luteinizing hormone (LH) is primarily concerned with the endocrine activities of the ovaries or testes, and follicle stimulating hormone (FSH) with their gametogenic roles. Production of both LH and FSH is, in turn, under the control of an oligopeptide releasing factor produced by the hypothalamus, and known as luteinizing hormone releasing hormone (LHRH; alias LH-releasing factor, LRF). The whole system, in both sexes, is controlled by a variety of feedback relationships, dominated by the actions of steroid hormones on hypothalamus and pituitary.

Antibodies against gonadal steroids would certainly impair fertility. Steroids are not, on their own, good antigens but coupled to suitable carrier molecules they can, as haptens (p. 4), evoke specific antibody formation. A major drawback to immunization against gonadal steroids, as a potential means of contraception in humans, is that the physiological effects of steroids are widespread and diverse. In particular, their sustaining effects on secondary sexual characteristics and on sexual behaviour would probably be missed were they to be suppressed immunologically. Immunization against steroids—functionally equivalent to castration or ovariectomy—does not therefore seem a fruitful line of approach.

LH is potentially an effective immunogen, and is not species-specific among mammals. As an antigen, it does cross-react to some extent with FSH and with thyroid-stimulating hormone (TSH), with both of which it has a considerable part of its chemical structure in common (Moudgal *et al.*, 1974).

Immunization of female macaque monkeys with sheep LH caused a suppression of ovulation followed by rapid recovery. Again, however, practical applications would be limited by the secondary effects of LH deprivation on the endocrine function of the gonads, since LH stimulates androgen and oestrogen production in, respectively, men and women. For the same reasons as previously, this would not seem likely to prove popular.

There might, however, be some practical application of anti-LH immunization in manipulating the reproduction of domestic animals. In cattle, for example, traditional methods of castration of bull calves when young mean forgoing the stimulatory effects of testosterone on muscle growth (that is, on production of usable meat); while castration in late adolescence is a major operation which probably causes a check in growth. An escape from this dilemma might be to achieve functional castration by immunological means. This could be done whenever required, with relatively little trauma and in consequence less likelihood of a growth check.

FSH is more directly involved with the gametogenic function of the gonads. Like LH, it can be an effective immunogen, and conveniently lacks species-specificity. Following the active immunization of male rhesus monkeys with purified sheep FSH, anti-FSH antibodies duly appeared and spermatogenesis was successfully suppressed for over two years. LH, testosterone and sexual behaviour were not significantly affected, compared with control animals, but the testis volume of the immunized animals fell, and mean sperm counts in their ejaculates dwindled to less than one-third those of the controls. Active immunization against FSH (and passive immunization likewise) therefore achieves selective removal of the hormone and a decline in spermatogenesis with little or no concomitant effect on other parts of the reproductive system. Fertility was not directly estimated, nor was the question of recovery from immunization investigated. Nevertheless, immunization against FSH does seem a possible line to pursue in the search for immunological methods of contraception (Wickings and Nieschlag, 1980).

Coupled to a suitable carrier, LHRH can also be immunogenic, despite its small molecular size. Again, it lacks species-specificity. Active and passive immunization of female macaque monkeys, marmosets, rabbits, mice and dogs reversibly suppresses ovulation, hence fertility. As LHRH determines the output of LH as well as of FSH, anti-LHRH antibodies affect the endocrine functions of the gonads and, through this, secondary sexual characteristics. For this reason, the application of anti-LHRH immunization in humans seems unlikely, but there may well prove to be useful applications in stockrearing (Fraser, 1980; Hodges and Hearn, 1979; Talwar *et al.*, 1980).

When a human embryo enters the uterus and begins to implant in the endometrium, it informs the corpus luteum of its arrival by the release of a glycoprotein hormone, manufactured by the trophoblast, and known as human chorionic gonadotrophin (HCG) (p. 54). (Homologous hormones are known in other primates, and may occur also in mammals other than primates.) HCG is therefore characteristic of pregnancy, and eliminating it by immunological methods should prevent the maternal recognition of pregnancy without affecting other physiological mechanisms; even ovulation and menstrual cycles should proceed without interruption.

Women are usually immunologically tolerant of HCG, and fail to produce antibodies against it. There are exceptions to this (Wass *et al.*, 1975), and a slight correlation of circulating anti-HCG antibody with infertility and a propensity for spontaneous abortion (also found with anti-HCS antibodies; see p. 54) has been reported; but such exceptions are rare. As HCG has been detected on some tumour cells (McManus *et al.*, 1976), and as certain bacteria synthesize a substance with the biological and immunological properties of the hormone (Cohen and Strampp, 1976), the breakdown of tolerance in women may not represent a response to trophoblast HCG, but to a version of it from an atypical source.

For deliberate immunization against HCG to be considered as a practical form of contraception, it is necessary first to persuade the immune system to react against a previously acceptable antigen. As with the hormones previously discussed, an obvious way to do this is to present the antigen together with a suitable adjuvant. This does indeed result in the production of anti-HCG antibodies. Unfortunately, these are liable to cross-react with the anterior pituitary hormones LH, FSH and TSH (thyroid-stimulating hormone). Induction of antibodies which react with the first two of these has effects which have already been discussed; interference with TSH could be disastrous. A more selective approach is called for.

HCG and the three anterior pituitary hormones with which anti-HCG antibody is likely to cross-react have similar molecular structures, consisting of two polypeptide chains. These are known as α and β chains; the α-chains of all four hormones are virtually identical, the hormonal individuality being determined by different β-chains. Immunological cross-reactivity stems largely from antigenic determinants on the common α-chains; the antigenic differences between the hormones must depend on differences between β-chains.

The β-subunit of HCG (HCG-β) comprises 145 amino acids and that of human LH 115. Counting from the N-terminal end of the β-chain, the first 110 amino acids show extensive sequence homology between HCG-β and LH-β; the amino acids in 94 positions (85%) are identical. Between positions 111 and 115 (115 being the final position of LH-β) only one amino acid is common to both. Thereafter HCG-β has a run of 30 amino acids which are simply lacking from the LH β-chain.

This suggests two strategies for the development of an anti-HCG vaccine which does not cross-react with LH: immunization against whole β-chain, hoping for minimal cross-reactivity with LH; or immunizing against peptides which correspond to the C-terminal amino acid sequence unique to HCG-β. In either case, coupling of the antigen to appropriate carrier molecules, and injection in adjuvant, are likely to prove essential.

Enzymic digestion of HCG-β results in a number of peptides of varying length. Those derived from amongst the 35 distinctive amino acids at the C-terminal end are immunogenic but, whereas the antibodies produced do react with HCG, significant ability to neutralize its biological activity is present only when the immunizing peptide is of at least 35 amino acids in length. Accordingly, large peptides have been used in attempts to immunize against pregnancy. Because of the low yield of suitable peptides from the direct enzymic breakdown of HCG-β, these have in practice been replaced by their synthesized equivalents; identical in amino acid sequence, but lacking the attached carbohydrate characteristic of the naturally-occurring material. Synthetic peptides, coupled to a carrier molecule such as tetanus toxoid (TT), produce significant amounts of anti-peptide antibody which reacts with HCG and not with LH.

When this antibody, prepared in baboons, was injected into a female baboon on the 18th day of her pregnancy, the passive immunity to HCG so conferred led to a rapid termination of pregnancy and resumption of menstrual cycles. When female baboons were actively immunized with the 35-amino acid peptide coupled to TT, and then mated with male baboons of proven fertility, the conception rate fell to 13%, compared with 81% for untreated controls and 75% for control females treated with TT alone. Although a contraceptive failure rate of 13% is too high to be tolerated in a human contraceptive technique, it must be remembered that immunizing baboons against human chorionic gonadotrophin is likely to be less effective than using baboon chorionic gonadotrophin as a source of antigen. With homologous hormone as antigen, better protection from pregnancy might result.

Both active and passive immunization against whole HCG in adjuvant has been attempted in marmosets. Passive immunization by the infusion of

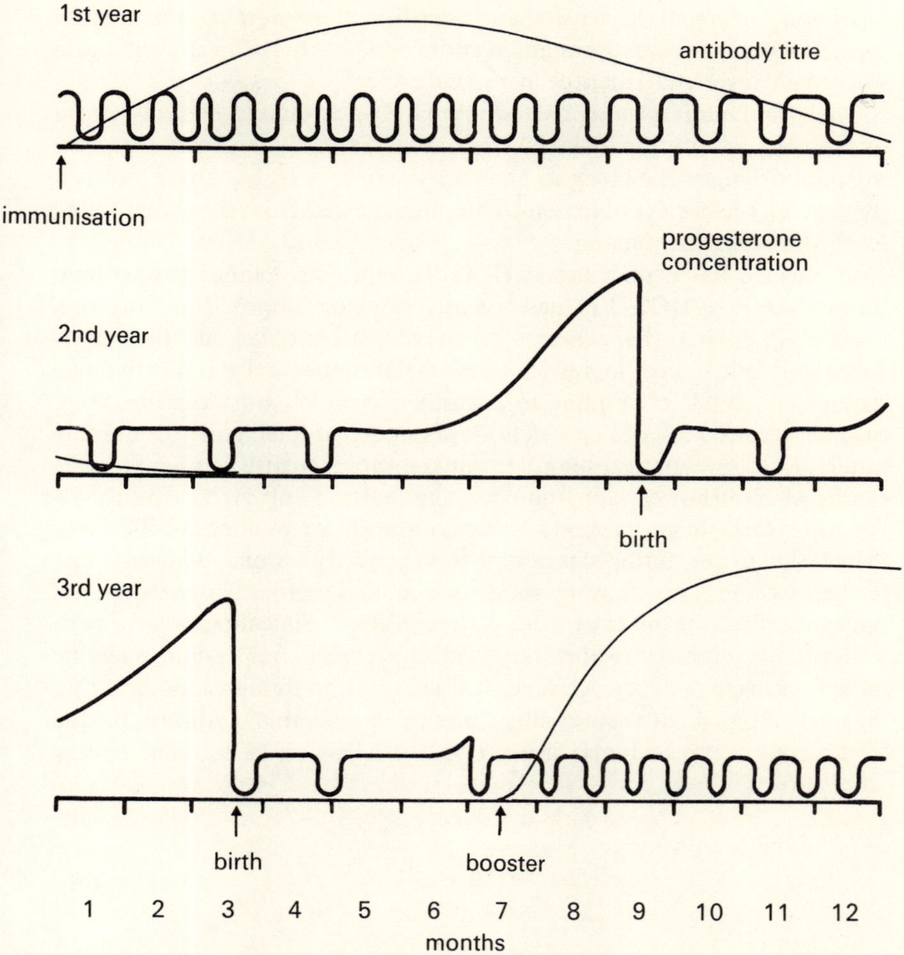

Figure 8.3 The breeding history, over three years, of a female marmoset (*Callithrix jacchus*) immunized against HCG-β. When antibody concentrations were high, pregnancy did not last beyond the length of the normal menstrual cycle, but as levels declined cycles became extended and recurrent abortions were noted. Pregnancies which went beyond 12 weeks proceeded to term. Such births were usually followed by a return to recurrent abortions. A booster injection $2\frac{1}{2}$ years after the initial immunization caused a repetition of this pattern. Fertilization occurred normally and the uterus showed no signs of abnormality (redrawn from Hearn, 1980).

(marmoset) antibodies terminated pregnancy up to six weeks from conception. Figure 8.3 summarizes the history of one female marmoset actively immunized against HCG-β. While antibody titres were high, menstrual cycles continued normally, as indicated by frequently recurring rises in luteal progesterone. As the anti-HCG-β levels fell, a series of prolonged cycles developed, in which conception was followed by early abortion. Interspersed with these infertile cycles were two successful pregnancies. A booster dose caused antibody levels to rise again, and again the female reverted to infertile menstrual cycles.

Active immunization against whole HCG-β is therefore effective as a means of preventing pregnancy, as long as antibody concentrations are maintained high. The block to pregnancy is reversible, but at the cost of a high incidence of early abortions. This might detract from the acceptability of the technique in humans.

A vaccine based on purified HCG-β coupled to tetanus toxoid (and known as Pr-β-HCG-TT) has recently been developed. Immunization against Pr-β-HCG-TT prevents pregnancy in baboons, and the results were sufficiently encouraging for clinical trials to be carried out on women volunteers. While continuing to use other forms of contraception, these women received a series of injections of antigen at fortnightly or monthly intervals. Although the response to antigen varied greatly between women, many women showed high and persistent levels of anti-HCG-β antibody. In some cases, high antibody levels continued for as long as 500 days. Menstrual cycles during this period were perfectly normal, ovulation and corpus luteum function were not disturbed, and there was no evidence of any side-effects of the treatment. Although these clinical trials have been criticized as premature, and there is disagreement over certain scientific and legal aspects of the vaccine, it does seem as though immunization against HCG might eventually provide a valuable addition to the contraceptive repertoire (Hearn, 1979, 1980; Talwar, 1979, 1980; Talwar *et al.*, 1980).

BIBLIOGRAPHY

Abedin, M. and Kirkpatrick, C. H. (1980) Immunosuppressive activity of cord blood leukocytes. *Pediatrics* **66**: 405–10.

Ablin, R. J., Soanes, W. A., Bronson, P. and Witebsky, E. (1970) Precipitating antigens of the normal human prostate. *J. Reprod. Fertil.* **22**: 573–4.

Adcock, E. W., Teasdale, F., August, C. S., Cox, S., Meschia, G., Battaglia, F. C. and Naughton, M. A. (1973) Human chorionic gonadotrophin: its possible role in maternal lymphocyte suppression. *Science* **181**: 845–7.

Adinolfi, M. (1975) The human placenta as a filter for cells and plasma proteins. *See* Edwards *et al.* (1975), pp. 191–215.

Aitken, R. J. and Richardson, D. W. (1980) Immunization against zona pellucida antigens. *See* Hearn (1980), pp. 173–201.

Aitken, R. J., Rudak, E. A., Richardson, D. W., Dor, J., Djahanbahkch, O. and Templeton, A. A. (1981) The influence of anti-zona and anti-sperm antibodies on sperm–egg interactions. *J. Reprod. Fertil.* **62**: 597–606.

Alexander, P. (1974) Escape from immune destruction by the host through shedding of surface antigen: is this a characteristic shared by malignant and embryonic cells? *Cancer Res.* **34**: 2077–82.

Allen, G. J. and Bourne, F. J. (1978) Interaction of immunoglobulin fragments with the mammalian sperm acrosome. *J. exp. Zool.* **203**: 271–6.

Amoroso, E. C. and Perry, J. S. (1975) The existence during gestation of an immunological barrier between maternal and foetal tissues. *Phil. Trans. roy. Soc. B* **271**: 343–61.

Amos, D. B. and Kostyu, D. D. (1980) HLA—a central immunological agency of man. *Adv. Human Genet.* **10**: 137–208.

Andrada, J. A., Comini, E. and Vilar, O. (1979) Immune response to spermatozoa, in *Immunopathology*, 6th Int. Congr. Immunol., Niagara Falls, N.Y., 1978, F. Milgrom and B. Albini (eds.), Karger, Basel, pp. 78–84.

Artzt, K. and Bennett, D. (1975) Analogies between embryonic (T/t) antigens and adult major histocompatibility (H-2) antigens. *Nature* **256**: 545–7.

Austin, C. R. and Short, R. V. (eds.) (1972*a*) *Reproduction in Mammals.* I: Germ cells and fertilization. Cambridge University Press.

Austin, C. R. and Short, R. V. (eds.) (1972*b*) *Reproduction in Mammals.* II: Embryonic and fetal development. Cambridge University Press.

Austin, C. R. and Short, R. V. (eds.) (1972*c*) *Reproduction in Mammals.* III: Hormones in reproduction. Cambridge University Press.

Austin, C. R. and Short, R. V. (eds.) (1972*d*) *Reproduction in Mammals.* IV: Reproductive patterns. Cambridge University Press.

Austin, C. R. and Short, R. V. (eds.) (1979) *Reproduction in Mammals.* VII: Mechanisms of hormone action. Cambridge University Press.

Bagshawe, K. D. and Lawler, S. (1975) The immunogenicity of the placenta and trophoblast. *See* Edwards *et al.* (1975), pp. 171–91.

Baines, M. G., Speers, E. A., Pross, H. and Millar, K. G. (1976) Characteristics of the maternal lymphoid response of mice to paternal strain antigens induced by homologous pregnancy. *Immunol.* **31**: 363–9.

Baines, M. G., Millar, K. G. and Pross, H. F. (1980) Allograft enhancement during normal murine pregnancy. *J. reprod. Immunol.* **2**: 141–9.

Balkwill, F. R. and Hogg, N. (1979) Characterization of human breast milk macrophages cytostatic for human cell lines. *J. Immunol.* **123**: 1451–6.

Balls, M. and Wild, A. E. (eds.) (1975) *The Early Development of Mammals.* Cambridge University Press.

Barg, M., Burton, R. C., Smith, J. A., Luckenbach, G. A., Decker, J. and Mitchell, G. F. (1978) Effects of placental tissue on immunological responses. *Clin. exp. Immunol.* **34**: 441–8.

Barnstaple, C. J. and Bodmer, W. F. (1978) Immunology and the fetus. *Lancet* 1978/1: 326.

Beer, A. E. (1979) The paradox of feto-placental units as allografts. *See* Talwar (1979), pp. 501–514.

Beer, A. E. (1980) Immunology of reproduction and embryonic development. *See* Fougereau and Dausset (1980), pp. 1137–45.

Beer, A. E. and Billingham, R. E. (1973) Maternally acquired runt disease. *Science* **179**:240–3.

Beer, A. E. and Billingham, R. E. (1974*a*) Host responses to intra-uterine tissue, cellular and fetal allografts. *J. Reprod. Fertil. Suppl.* **21**: 59–88.

Beer, A. E. and Billingham, R. E. (1974*b*) The embryo as a transplant. *Sci. Amer.* **230**: 36–46.

Beer, A. E. and Billingham, R. E. (1976) *The Immunobiology of Mammalian Reproduction.* Prentice-Hall.

Beer, A. E. and Billingham, R. E. (1978) Immunoregulatory aspects of pregnancy. *Fedn. Proc.* **37**: 2374–8.

Beer, A. E. and Billingham, R. E. (1979) Maternal immunological recognition mechanisms during pregnancy, in Ciba Fndn. Symp. 64 (N.S.): *Maternal Recognition of Pregnancy*, Excerpta Medica pp. 293–322.

Beer, A. E. Billingham, R. E. and Yang, S. L. (1972) Further evidence concerning the autoantigenic status of the trophoblast. *J. exp. Med.* **135**: 1177–84.

Beer, A. E. and Neaves, W. B. (1978) Antigenic status of semen from the viewpoints of the female and male. *Fertil. Steril.* **29**: 3–22.

Bell, E. B. (1969) Immunological control of fertility in mouse: a comparison of systemic and intravaginal immunization. *J. Reprod. Fertil.* **18**: 183–92.

Bell, S. C. and Billington, W. D. (1980) Major anti-paternal alloantibody induced by murine pregnancy is non-complement-fixing IgGl. *Nature* **288**: 387–8.

Beller, F. K. and Schumacher, G. F. B. (eds.) (1980) *The Biology of the Fluids of the Female Genital Tract.* Elsevier/North-Holland.

Bennett, D. (1975) The T-locus of the mouse. *Cell* **6**: 441–54.

Bernard, O. (1977) Possible protecting role of maternal immunoglobulins on embryonic development in mammals. *Immunogenet.* **5**: 1–15.

Bernard, O., Ripoche, M-A. and Bennett, D. (1977) Distribution of maternal immunoglobulins in the mouse uterus and embryo in the days after implantation. *J. exp. Med.* **145**: 58–75.

Bienenstock, J. and Befus, A. D. (1980) Mucosal immunology. *Immunol.* **41**: 249–70.

Billingham, R. E. and Silvers, W. K. (1964) Studies on homografts of foetal and infant skin and further observations on the anomalous properties of pouch skin grafts in hamsters. *Proc. roy. Soc. B* **161**: 168–90.

Billington, W. D. (1969) Trophoblast extensions from the placenta. *Proc. roy. Soc. Med.* **63**: 57–9.

Billington, W. D. (1975) Organization, ultrastructure and histochemistry of the placenta: immunological considerations. *See* Edwards *et al.* (1975), pp. 67–85.

Billington, W. D., Jenkinson, E. J., Searle, R. F. and Sellens, M. H. (1977) Alloantigen expression during early embryogenesis and placental ontogeny in the mouse: immunoperoxidase and mixed hemadsorption studies. *Transpl. Proc.* **9**: 1371–7.

Billington, W. D., Kirby, D. R. S., Owen, J. J. T., Ritter, M. A., Burtonshaw, M. D., Evans, E. P., Ford, C. E., Gauld, I. K. and McLaren, A. (1969) Placental barrier to maternal cells. *Nature* **224**: 704–6.

Birkeland, S. A. and Kristoffersen, K. (1980*a*) A longitudinal study of B lymphocyte responses during normal human pregnancy. *J. reprod. Immunol.* **2**: 23–7.

Birkeland, S. A. and Kristoffersen, K. (1980*b*) The fetus as an allograft: a longitudinal study of normal human pregnancies studied with mixed lymphocyte cultures between mother-father and mother-child. *Scand. J. Immunol.* **11**: 311–9.

Birkeland, S. A. and Kristoffersen, K. (1980*c*) Lymphocyte transformation with mitogens and antigens during normal human pregnancy: a longitudinal study. *Scand. J. Immunol.* **11**: 321–5.

Bissenden, J. C., Ling, N. R. and Mackintosh, P. (1980) Suppression of mixed lymphocyte reactions by pregnancy serum. *Clin. exp. Immunol.* **39**: 195–202.

Boettcher, B. (1968) Correlation between human ABO blood group antigens in seminal plasma and on seminal spermatozoa. *J. Reprod. Fertil.* **16**: 49–54.

Boettcher, B. (1974) The molecular nature of spermagglutinins and sperm antibodies in human sera. *J. Reprod. Fertil. Suppl.* **21**: 151–67.

Boettcher, B. (ed.) (1977) *Immunological Influences on Human Fertility.* Academic Press.

Borland, R., Loke, Y. W. and Wilson, D. (1975) Immunological privilege resulting from endocrine activity of trophoblast *in vivo. See* Edwards *et al.* (1975), pp. 157–69.

Boshier, D. P. (1968) The relationship between genotype and reproductive performance before parturition in mice. *J. Reprod. Fertil.* **15**: 427–35.

Brambell, F. W. R. (1970) *The Transmission of Passive Immunity from Mother to Young.* Elsevier/North-Holland.

Brown, P. C., Glynn, L. E. and Holborrow, E. J. (1967) The dual necessity for delayed hypersensitivity and circulating antibody in the pathogenesis of experimental allergic orchitis in guinea-pigs. *Immunol.* **13**: 307–14.

Bulmer, R. and Hancock, K. W. (1977) Depletion of circulating T lymphocytes in pregnancy. *Clin. exp. Immunol.* **28**: 302–5.

Burnet, F. M. (ed.) (1976) *Immunology* (Readings from *Scientific American*) W. H. Freeman.

Carlson, G. A. and Wegmann, T. G. (1978) Paternal-strain antigen excess in semi-allogeneic pregnancy. *Transpl. Proc.* **10**: 403–7.

Carter, J. (1978) The expression of surface antigens on three trophoblastic tissues in the mouse. *J. Reprod. Fertil.* **54**: 433–9.

Castro, J. E., Hunt, R., Lance, E. M. and Medawar, P. B. (1974) Implications of the fetal antigen theory for fetal transplantation. *Cancer Res.* **34**: 2055–60.

Catty, D. and Lowe, J. A. (1976) Effect of maternal anti-immunoglobulin (anti-allotypic) antibody on the synthesis of immunoglobulin in the neonatal rabbit. *See* Hemmings (1976), pp. 261–71.

Chandra, R. K. (1976) Levels of IgG subclasses, IgA, IgM, and tetanus antitoxin in paired maternal and foetal sera: findings in healthy pregnancy and placental insufficiency. *See* Hemmings (1976), pp. 77–90.

Chatterjee-Hasrouni, S. and Lala, P. K. (1979) Localization of H-2 antigens on mouse trophoblast cells. *J. exp. Med.* **149**: 1233–53.

Chatterjee-Hasrouni, S., Santer, V. and Lala, P. K. (1980) Characterization of maternal small lymphocyte subsets during allogeneic pregnancy in the mouse. *Cell. Immunol.* **50**: 290–304.

Chaouat, G. and Voisin, G. A. (1979) Regulatory T cell subpopulations in pregnancy. I. Evidence for suppressive activity of the early phase of MLR. *J. Immunol.* **122**: 1383–8.

Chaouat, G. and Voisin, G. A. (1980): Regulatory T cell subpopulations in pregnancy. II. Evidence for suppressive activity of the late phase of MLR. *Immunol.* **39**: 239–48.

Chaouat, G., Voisin, G. A., Escalier, D. and Robert, P. (1979) Facilitation reaction (enhancing antibodies and suppressor cells) and rejection reaction (sensitized cells) from the mother to paternal antigens of the conceptus. *Clin. exp. Immunol.* **35**: 13–24.

Chardonnens, X. and Jeannet, M. (1980) Immunobiology of pregnancy: evidence for a fetal immune response against the mother (?). *Tissue Antigens* **15**: 401–6.

Chen, C. and Simons, M. J. (1977) Modified seminal plasma antigens and subfertility. *See* Boettcher (1977), pp. 263–70.

Cinader, B. and de Weck, A. (eds.) (1976) *Immunological Response of the Female Reproductive Tract*. W.H.O./Scriptor, Copenhagen.

Clark, D. A. and McDermott, M. R. (1978) Impairment of host versus graft reaction in pregnant mice. I. Suppression of cytotoxic T cell generation in lymph nodes draining the uterus. *J. Immunol.* **121**: 1389–93.

Clark, D. A., McDermott, M. R. and Szewczuk, M. R. (1980) Impairment of host-versus-graft reaction in pregnant mice. II. Selective suppression of cytotoxic T-cell generation correlates with soluble suppressor activity and with successful allogeneic pregnancy. *Cell. Immunol.* **52**: 106–18.

Clarke, F. M., Morton, H., Rolfe, B. E. and Clunie, G. J. A. (1980) Partial characterisation of early pregnancy factor in the sheep. *J. reprod. Immunol.* **2**: 151–62.

Clemens, L. E., Siiteri, P. K. and Stites, D. P. (1979) Mechanism of immunosuppression of progesterone on maternal lymphocyte activation during pregnancy. *J. Immunol.* **122**: 1978–85.

Cohen, H. and Strampp, A. (1976) Bacterial synthesis of substance similar to human chorionic gonadotrophin. *Proc. Soc. exp. Biol. Med.* **152**: 408–10.

Cohen, J. (1969) Why so many sperms? An essay on the arithmetic of reproduction. *Sci. Prog.* **57**: 23–41.

Cohen, J. and Gregson, S. H. (1978) Antibodies and sperm survival in the female genital tract. *See* Cohen and Hendry (1978), pp. 17–29.

Cohen, J. and Hendry, W. F. (eds.) (1978) *Spermatozoa, Antibodies and Infertility*. Blackwell.

Cohen, J. and Tyler, K. R. (1980) Sperm populations in the female genital tract of the rabbit. *J. Reprod. Fertil.* **60**: 213–8.

Cohen, S. (ed.) (1979) *Biology of the Lymphokines*. Academic Press.

Coligan, J. E., Kindt, T. J., Uehara, H., Martinko, J. and Nathenson, S. G. (1981) Primary structure of a murine transplantation antigen. *Nature* **291**:35–9.

Collins, G. D., Chrest, F. J. and Adler, W. H. (1980) Maternal cell traffic in allogeneic embryos. *J. reprod. Immunol.* **2**: 163–72.

Contractor, S. F. and Davies, H. (1973) Effect of human chorionic somatomammotrophin and human chorionic gonadotrophin on phytohaemagglutinin-induced lymphocyte transformation. *Nature New Biol.* **243**: 284–6.

Cooper, D. W. (1980) Immunological relationships between mother and conceptus in man. *See* Hearn (1980), pp. 33–61.

Cooper, D. W. and Aitken, R. J. (1981) Failure to detect altered rosette inhibition titres in human pregnancy serum. *J. Reprod. Fertil.* **61**: 241–5.

Crago, S. S., Prince, S. J., Pretlow, T. G., McGhee, J. R. and Mestecky, J. (1979) Human colostral cells. I. Separation and characterization. *Clin. exp. Immunol.* **38**: 585–97.

Currie, G. A. (1974) *Cancer and the Immune Response*. Edward Arnold.

d'Almeida, M. and Voisin, G. A. (1979) Resistance of female guinea pig fertility to efficient iso-immunization with spermatozoa autoantigens. *J. reprod. Immunol.* **1**: 237–47.

Davies, D. A. L. and Staines, N. A. (1976) A cardinal role for I-region antigens (Ia) in immunological enhancement, and the clinical implications. *Transplant. Rev.* **30**: 18–39.

de Fazio, S. R. and Ketchel, M. M. (1972) The occurrence of seminal plasma antigens in the tissues of women. *J. Reprod. Fertil.* **30**: 125–131.

Dickman, W. J. and Cauchi, M. N. (1978) Lymphocyte-induced stimulation of human chorionic gonadotrophin production by trophoblastic cells *in vitro*. *Nature* **271**: 377–8.

Diczfalusy, E. (ed.) (1974) *Immunological Approaches to Fertility Control*. Karolinska Institutet, Stockholm.

Dooher, G. B., Artzt, K., Bennett, D. and Hurtenbach, U. (1981) Observations on autoimmune orchitis in sterile mice carrying a recessive lethal mutation at the T/t complex. *J. Reprod. Fertil.* **62**: 505–11.

Dorsman, B. G., Tumboh-Oeri, A. G. and Roberts, T. K. (1978) Detection of cell-mediated immunity to spermatozoa in mice and man by the leucocyte adherence-inhibition test. *J. Reprod. Fertil.* **53**: 277–83.

Doughty, R. W. and Gelsthorpe, K. (1974) An initial investigation of lymphocyte antibody activity through pregnancy and in eluates prepared from placental material. *Tissue Antigens* **4**: 291–6.

Doughty, R. W. and Gelsthorpe, K. (1976) Some parameters of lymphocyte antibody activity through pregnancy and further eluates of placental material. *Tissue Antigens* **8**: 43–8.

Ducibella, T. (1980) Divalent antibodies to mouse embryonal carcinoma cells inhibit compaction in the mouse embryo. *Developmental Biol.* **79**: 356–66.

Dym, M. and Romrell, L. J. (1975) Intraepithelial lymphocytes in the male reproductive tract of rats and rhesus monkeys. *J. Reprod. Fertil.* **42**: 1–7.

Edidin, M. and Johnson, M. H. (eds.) (1977) *Immunobiology of Gametes.* Cambridge University Press.

Edwards, R. G. (1960) Antigenicity of rabbit semen, bull semen and egg yolk after intravaginal or intramuscular injections into female rabbits. *J. Reprod. Fertil.* **1**: 385–401.

Edwards, R. G., Ferguson, L. C. and Coombs, R. R. A. (1964) Blood group antigens on human spermatozoa. *J. Reprod. Fertil.* **7**: 153–61.

Edwards, R. G., Howe, C. W. S. and Johnson, M. H. (eds.) (1975) *Immunobiology of Trophoblast.* Cambridge Univ. Press.

Emerson, S. G., Murphy, D. B. and Cone, R. E. (1980) Selective turnover and shedding of H-2K and H-2D antigens is controlled by the major histocompatibility complex. Implications for H-2 restricted recognition. *J. exp. Med.* **152**: 783–95.

Fabris, N. (1973) Immunological reactivity during pregnancy in the mouse. *Experientia* **29**: 610–2.

Fabris, N., Piantanelli, L. and Muzzioli, M. (1977) Differential effect of pregnancy or gestagens on humoral and cell-mediated immunity. *Clin. exp. Immunol.* **28**: 306–14.

Faulk, W. P. (1980) Immunology of the maternofoetal relationship. *See* Fougereau and Dausset (1980), pp. 1093–1117.

Faulk, W. P. and Galbraith, G. M. P. (1979) Trophoblast transferrin and transferrin receptors in the host-parasite relationship of human pregnancy. *Proc. roy. Soc. B* **204**: 83–97.

Faulk, W. P., Jeannet, W. D., Creighton, W. D. and Carbonara, A. (1974) Immunological studies of the human placenta. Characterization of immunoglobulins on trophoblastic basement membrane. *J. clin. Invest.* **54**: 1011–9.

Faulk, W. P. and Johnson, P. M. (1977) Immunological studies of human placentae: identification and distribution of proteins in mature chorionic villi. *Clin. exp. Immunol.* **27**: 365–75.

Faulk, W. P., Sanderson, A. and Temple, A. (1977) Distribution of MHC antigens in human placentae. *Transpl. Proc.* **9**: 1379–84.

Faulk, W. P. and Temple, A. (1976) Distribution of β_2 microglobulin and HLA in chorionic villi of human placentae. *Nature* **262**: 799–802.

Faulk, W. P., Temple, A., Lovins, R. and Smith, N. C. (1978) Antigens of human trophoblasts: a working hypothesis for their role in normal and abnormal pregnancies. *Proc. nat. Acad. Sci.* **75**: 1947–51.

Fauve, R. M., Hevin, B., Jacob, H., Gaillard, J. A. and Jacob, F. (1974) Antiinflammatory effects of murine malignant cells. *Proc. nat. Acad. Sci.* **71**: 4052–6.

Fawcett, D. W. (1975) Ultrastructure and function of the Sertoli cell, in *Handbook of Physiology*, Section 7, *Endocrinology*, vol. V. Male Reproductive System, Greep, R. O. and Astwood, E. B. (eds.) Amer. Physiol. Soc., Washington, D.C., pp. 21–55.

Fellous, M. and Dausset, J. (1970) Probable haploid expression of HL-A antigens on human spermatozoa. *Nature* **225**: 191–3.

Fellous, M., Erickson, R. P., Gachelin, G. and Jacob, F. (1976) The time of appearance of Ia antigens during spermatogenesis in the mouse. *Transpl.* **22**: 440–4.

Ferguson, F. G. and Palm, J. (1977) Reactivity of rat placental cells with alloantisera. *J. Embryol. exp. Morphol.* **39**: 195–202.

Ferrone, S., Mickey, M. R., Terasaki, P. I., Reisfeld, R. A. and Pellegrino, M. A. (1976) Humoral sensitization in parous women; cytotoxic antibodies to non-HL-A antigens. *Transpl.* **22**: 61–8.

Finn, R., Hill, C. A. S., Govan, A. J., Ralfs, I. G., Gurney, F. J. and Denye, V. (1972) Immunological responses in pregnancy and survival of fetal homograft. *Brit. med. J.* 1972/3: 150–2.

Fjällbrant, B. (1968*a*) Sperm agglutinins in sterile and fertile men. *Acta obstet. gynecol. scand.* **47**: 89–101.

Fjällbrant, B. (1968*b*) Interrelation between high levels of sperm antibodies, reduced penetration of cervical mucus by spermatozoa, and sterility in men. *Acta obstet. gynecol. scand.* **47**: 102–18.

Fjällbrant, B. (1975): Autoimmune human sperm antibodies and age in males. *J. Reprod. Fertil.* **42**: 145–8.

Fougereau, M. and Dausset, J. (eds.) (1980) *Immunology 80* (Progress in Immunology IV) Academic Press.

Fowler, A. K., Reed, C. D. and Giron, D. J. (1980) Identification of an interferon in murine placentas. *Nature* **286**: 266–7.

Franklin, R. R. and Dukes, C. D. (1964) Antispermatozoal antibody and unexplained infertility. *Amer. J. Obstet. Gynecol.* **98**: 6–9.

Fraser, H. M. (1980) Inhibition of reproductive function by antibodies to luteinizing hormone releasing hormone. *See* Hearn (1980), pp. 143–171.

Friend, D. S. and Gilula, N. B. (1972) Variations in the tight and gap junctions in mammalian tissues. *J. cell Biol.* **53**: 758–76.

Fuchs, T., Hammarström, L., Smith, C. I. and Brundin, J. (1980) *In vitro* induction of murine suppressor T-cells by HCG. *Acta obstet. gynecol. scand.* **59**: 355–60.

Fudenberg, H. H. and Fudenberg, B. R. (1964) Antibody to hereditary human gamma-globulin (Gm) factor resulting from maternal-fetal incompatibility. *Science* **145**: 170–1.

Fujisaki, S., Mori, N., Sasaki, T. and Maeyama, N. (1979) Cell-mediated immunity in pregnancy. Changes in lymphocyte reactivity during pregnancy and postpartum. *Microbiol. Immunol.* **23**: 899–907.

Gachelin, G., Fellous, M., Guénet, J-L. and Jacob, F. (1976) Developmental expression of an early embryonic antigen common to mouse spermatozoa and cleavage embryos, and to human spermatozoa: its expression during spermatogenesis. *Developmental Biol.* **50**: 310–20.

Gaillard, M. C. and Liegeois, A. (1980) Microchimerism: an experimental model and a physiological fact during pregnancy. *See* Touraine *et al.* (1980), pp. 137–42.

Gardner, R. L. (1975) Origins and properties of trophoblast. *See* Edwards *et al.* (1975), pp. 43–65.

Gleicher, N., Beers, P., Kerenyi, T. D., Cohen, C. J. and Gusberg, S. B. (1980) Leukocyte migration enhancement as an indicator of immunologic enhancement. I. Pregnancy. *Transpl.* **23**: 423–30.

Goldberg, E. (1974) Effects of immunization with LDH-X on fertility. *See* Diczfalusy (1974), pp. 202–222.

Goldberg, E. and Lerum, J. (1972) Pregnancy suppression by an antiserum to the sperm specific lactic dehydrogenase. *Science* **176**: 686–7.

Goldberg, E., Wheat, R. E., Powell, J. E. and Stevens, V. C. (1981) Reduction in fertility in female baboons immunized with lactic dehydrogenase-C4. *Fertil. Steril.* **35**: 214–7.

Goldstein, S. L., Anderson, R. G. W. and Brown, M. S. (1979) Coated pits, coated vesicles, and receptor-mediated endocytosis. *Nature* **279**: 679–85.

Golub, E. S. (1980) Know thyself: autoreactivity in the immune response. *Cell* **21**: 603–4.

Goodfellow, P. N., Levinson, J. R., Gable, R. J. and McDevitt, H. O. (1979) Analysis of anti-sperm sera for T/t locus specificity. *J. reprod. Immunol.* **1**:11–21.

Goodlin, R. C. and Herzenberg, L. A. (1964) Pregnancy-induced haemagglutinins to paternal H-2 antigens in multiparous mice. *Transpl.* **2**: 357–61.

Gottesman, S. R. and Stutman, O. (1978) Defect of cell-mediated immunity during pregnancy. *Fedn. Proc.* **37**: 1474.

Gundert, D., Merz, W. E., Hilgenfeldt, U. and Brossmer, R. (1975) Inability of highly purified preparations of human chorionic gonadotrophin to inhibit the PHA-induced stimulation of lymphocytes. *FEBS Lett.* **53**: 309–12.

Gusdon, J. P. (1976) Maternal immune responses in pregnancy. *See* Scott and Jones (1976), pp. 103–25.

Hagopian, A., Jackson, J. J., Carlo, D. J., Limjuco, G. A. and Eylar, E. H. (1975) Experimental allergic aspermatogenic orchitis. III. Isolation of spermatozoal glycoproteins and their role in allergic aspermatogenic orchitis. *J. Immunol.* **115**: 1731–43.

Hagopian, A., Limjuco, G. A., Jackson, J. J., Carlo, D. J. and Eylar, E. H. (1976) Experimental allergic aspermatogenic orchitis. IV. Chemical properties of spermatozoal glycoproteins isolated from guinea pig testis. *Biochim. biophys. Acta* **434**: 354–64.

Håkansson, S., Heyner, S., Sundqvist, K-G. and Bergström, S. (1975) The presence of paternal H-2 antigens on hybrid mouse blastocysts during experimental delay of implantation and the disappearance of these antigens after onset of implantation. *Int. J. Fertil.* **20**: 137–40.

Håkansson, S. and Sundqvist, K-G. (1975) Decreased antigenicity of mouse blastocysts after activation for implantation from experimental delay. *Transpl.* **19**: 479–84.

Halpern, B. N., Ky, T. and Robert, B. (1967) Clinical and immunological study of an exceptional case of reaginic type sensitization to human seminal fluid. *Immunol.* **12**: 247–58.

Hamilton, M. S. and Hellström, I. (1977) Altered immune response in pregnant mice. *Transpl.* **23**: 423–30.

Hamilton, M. S. and Hellström, I. (1978) Selection for histoincompatible progeny in mice. *Biol. Reprod.* **19**: 267–70.

Hamilton, M. S., Vitetta, E. S. and Beer, A. E. (1978) The effect of immunizing female mice to F9 teratocarcinoma cells on subsequent conception. *Fedn. Proc.* **37**: 1474.

Hammarström, L., Fuchs, T. and Smith, C. I. E. (1979) The immunodepressive effect of human glucoproteins and their possible role in the nonrejection process during pregnancy. *Acta obstet. gynecol. scand.* **58**: 417–22.

Hancock, R. J. T. (1978) Sperm antigens and sperm immunogenicity. *See* Cohen and Hendry (1978), pp. 1–9.

Hansen, K. B. and Hjort, T. (1971) Immunofluorescent studies on human spermatozoa. II. Characterization of spermatozoal antigens and their occurrence in spermatozoa from the male partners of infertile couples. *Clin. exp. Immunol.* **9**: 312–23.

Hearn, J. P. (ed.) (1980) *Immunological Aspects of Reproduction and Fertility Control.* MTP Press.

Hekman, A. and Rümke, P. (1969) The antigens of human seminal plasma, with special reference to lactoferrin as a spermatozoa-coating antigen. *Fertil. Steril.* **20**: 312–23.

Hellström, I., Hellström, K. E. and Allison, A. C. (1971) Neonatally induced allograft tolerance may be mediated by serum-borne factors. *Nature* **230**: 49–50.

Hemmings, W. A. (ed.) (1976) *Maternofoetal Transmission of Immunoglobulins.* (Proceedings of Symposium on Transmission of Immunoglobulins from Mother to Young.) Cambridge University Press.

Hemmings, W. A. (ed.) (1979) *Protein Transmission through Living Membranes.* Elsevier/North-Holland.

Herr, J. L. and Eddy, E. M. (1980) Detection of mouse sperm antigens by a surface labelling and immunoprecipitation approach. *Biol. Reprod.* **22**: 1263–74.

Heyner, S., Brinster, R. L. and Palm, J. (1969) Effect of allo-antibody on pre-implantation mouse embryos. *Nature* **222**: 783–4.

Heyner, S., Hunziker, R. D. and Zink, G. L. (1980) Differential expression of minor histocompatibility antigens on the surface of the mouse oocyte and preimplantation developmental stages. *J. reprod. Immunol.* **2**: 269–79.

Hill, C. A. S., Finn, R. and Denye, V. (1973) Depression of cellular immunity in pregnancy due to a serum factor. *Brit. med. J.* 1973/3: 513–4.

Hirahara, F., Gorai, I., Matsuzaki, Y., Sumiyoshi, Y. and Shiojima, Y. (1980) Cellular immunity in pregnancy: subpopulations of T lymphocytes bearing receptors for IgG and IgM in pregnant women. *Clin. exp. Immunol.* **41**: 353–7.

Hjort, T. (1977) Immunological capacity of the male genital tract. *See* Boettcher (1977), pp. 115–117.

Hodges, J. K. and Hearn, J. P. (1979) Long term suppression of fertility by immunisation against LHRH and its reversibility in female and male marmosets. *See* Talwar (1979), pp. 87–96.

Hogarth, P. J. (1978) *Biology of Reproduction.* Blackie, Glasgow and London.

Hojo, K., Hiramine, C. and Ishitaki, M. (1980) Lymphocyte proliferative response *in vitro* and its cellular dependency in guinea-pigs with experimental allergic orchitis. *J. Reprod. Fertil.* **59**: 113–23.

Holden, T. E. and Sherline, D. M. (1973) Bestiality, with sensitization and anaphylactic reaction. *Obstet. Gynecol.* **42**: 138–40.

Horne, C. H. W., Towler, C. M., Pugh-Humphreys, R. G. P., Thomson, A. W. and Bohn, H. (1976) Pregnancy-specific β_1-glycoprotein—a product of the syncytiotrophoblast. *Experientia* **32**: 1197–9.

Howard, J. (1980) MHC restriction, self-tolerance and the thymus. *Nature* **286**: 15–6.

Hurtenbach, U., Morgenstern, F. and Bennett, D. (1980) Induction of tolerance *in vitro* by autologous murine testicular cells. *J. exp. Med.* **151**: 827–38.

Inchley, C. J. (1981) *Immunobiology*. Edward Arnold.

Irvine, W. J. and Barnes, E. W. (1974) Addison's disease and autoimmune ovarian failure. *J. Reprod. Fertil. Suppl.* **21**: 1–31.

Isojima, S., Koyama, K. and Tsuchiya, K. (1974) The effect on fertility in women of circulating antibodies against human spermatozoa. *J. Reprod. Fertil. Suppl.* **21**: 125–50.

Isojima, S. and Li, T. S. (1968) Stepwise appearance of sperm-specific antigens in rats and their disappearance after fertilization. *Fertil. Steril.* **19**: 999–1008.

Ivanyi, P. (1978) Some aspects of the H-2 system, the major histocompatibility system in the mouse. *Proc. roy. Soc. B* **202**: 117–58.

Jackson, J. J., Hagopian, A., Carlo, D. J., Limjuco, G. A. and Eylar, E. H. (1975) Experimental allergic aspermatogenic orchitis. I. Isolation of a spermatozoal protein (AP1) which induces allergic aspermatogenic orchitis. *J. biol. Chem.* **250**: 6141–50.

Jacobs, B. B. and Uphoff, D. E. (1974) Immunologic modification: a basic survival mechanism. *Science* **185**: 582–7.

Jenkins, D. M. (1976) Pre-eclampsia/eclampsia (gestosis) and other pregnancy complications with possible immunologic basis. *See* Scott and Jones (1976), pp. 297–327.

Jenkins, D. M. and Good, S. (1972) Mixed lymphocyte reaction and placentation. *Nature New Biol.* **240**: 211–2.

Jenkinson, E. J. and Owen, V. (1980) Ontogeny and distribution of major histocompatibility complex (MHC) antigens on mouse placental trophoblast. *J. reprod. Immunol.* **2**: 173–81.

Jenkinson, E. J. and Searle, R. F. (1979) Ia antigen expression on the developing mouse embryo and placenta. *J. reprod. Immunol.* **1**: 3–10.

Johanson, R. A., Shaw, A. R. and Schlamowitz, M. (1980) The CH2 domain of rabbit IgG as the site of the receptor recognition unit for materno-fetal transport. *Fedn. Proc.* **39**: 481.

Johnson, M. H. (1972) The distribution of immunoglobulin and spermatozoal autoantigen in the genital tract of the male guinea pig: its relationship to autoallergic orchitis. *Fertil. Steril.* **23**: 383–92.

Johnson, M. H. (1975*a*) The macromolecular organization of membranes and its bearing on events leading up to fertilization. *J. Reprod. Fertil.* **44**: 167–84.

Johnson, M. H. (1975*b*) Antigens of the peri-implantation trophoblast. *See* Edwards *et al.* (1975), pp. 87–112.

Johnson, M. H. and Everitt, B. J. (1980) *Essential Reproduction*. Blackwell.

Johnson, P. M., Brown, P. J. and Faulk, W. P. (1980) Immunobiological aspects of the human placenta. *Oxford Rev. Reprod. Biol.* **2**: 1–40.

Johnson, P. M., Faulk, W. P. and Wang, A-C. (1976) Immunological studies of human placentae: subclass and fragment specificity of binding of aggregated IgG by placental endothelial cells. *Immunol.* **31**: 659–64.

Johnson, P. M., Natvig, J. B., Ystehede, U. A. and Faulk, W. P. (1977) Immunological studies of human placentae: the distribution and character of immunoglobulins in chorionic villi. *Clin. exp. Immunol.* **30**: 145–53.

Jones, W. R. (1976) Fetal and neonatal immunology. *See* Scott and Jones (1976), pp. 127–67.

Jones, W. R. (1980) Immunological factors in male and female infertility. *See* Hearn (1980), pp. 105–40.

Kadowaki, J. I., Thompson, R. I., Zuelzer, W. W., Woolley, P. V., Brough, A. J. and Gruber, D. (1965) XX/XY lymphoid chimaerism in congenital immunological deficiency syndrome with thymic alymphoplasia. *Lancet* 1965/2: 1152–6.

Kasakura, S. (1973) Is cortisol responsible for inhibition of MLC reactions by pregnancy plasma? *Nature* **246**: 496–7.

Katsh, S., Aguirre, A. and Katsh, G. F. (1968) Inactivation of sperm antigens by sera and tissues of the female reproductive tract. *Fertil. Steril.* **19**: 740–7.

Kelly, J. K. and Fox, H. (1979) The local immunological defence system of the human endometrium. *J. reprod. Immunol.* **1**: 39–45.

Kerék, G. and Afzelius, B. (1973) Demonstration of HL-A antigens, "species", and "semen"-specific antigens on human spermatozoa. *Int. J. Fertil.* **18**: 145–55.

Kille, J. W. and Goldberg, E. (1979) Female reproductive tract immunoglobulin responses to a purified sperm specific antigen (LDH-C_4). *Biol. Reprod.* **20**: 863–71.

Kirby, D. R. S. (1965) The improbable trophoblast. *New Sci.* **25**: 561–4.

Kirby, D. R. S. (1968) Immunological aspects of pregnancy. *Adv. reprod. Physiol.* **3**: 33–80.

Kirby, D. R. S. (1970) The egg and immunology. *Proc. roy. Soc. Med.* **63**: 59–61.

Kirkwood, K. J. and Billington, W. D. (1981) Expression of serologically detectable H-2 antigens on mid-gestation mouse embryonic tissues. *J. Embryol. exp. Morphol.* **61**: 207–19.

Kitzmiller, J. L. and Rocklin, R. E. (1980) Lack of suppression of lymphocyte migration inhibition factor production by estradiol, progesterone and human chorionic gonadotrophin. *J. reprod. Immunol.* **1**: 297–306.

Klein, J., Juretić, A., Saxevanis, C. N. and Nagy, Z. A. (1981) The traditional and a new version of the mouse H-2 complex. *Nature* **291**: 455–60.

Kolk, A. H. J. and Samuel, T. (1975) Isolation, chemical and immunological characterization of two strongly basic nuclear proteins from human spermatozoa. *Biochim. biophys. Acta* **393**: 307–19.

Kolodny, R. C., Koehler, B. C., Toro, G. and Masters, W. H. (1971) Spermagglutinating antibodies and infertility. *Obstet. Gynecol.* **38**: 576–82.

Koo, G. C., Stackpole, C. W., Boyse, E. A., Hämmerling, U. and Lardis, M. (1973) Topographical location of H-Y antigen on mouse spermatozoa by immunoelectron-microscopy. *Proc. nat. Acad. Sci.* **70**: 1502–5.

Kremer, J., Jager, S., Kuiken, J. and van Slochteren-Draaisma, T. (1978) Recent advances in diagnosis and treatment of infertility due to antisperm antibodies. *See* Cohen and Hendry (1978), pp. 117–27.

Krupen-Brown, K. and Wachtel, S. S. (1979) Cytotoxic and agglutinating H-Y antibodies in multiparous female mice. *Transpl.* **27**: 406–9.

Lau, H. L. and Linkins, S. E. (1976) Alpha-fetoprotein. *Amer. J. Obstet. Gynecol.* **124**: 533–49.

Le Bouteiller, P., Toullet, F. and Voisin, G. A. (1975) Ultrastructural lesions induced *in vitro* in guinea-pig spermatozoa by a specific autoantibody (anti-T) and complement. *Immunol.* **28**: 983–9.

Leiper, J. B. and Solomon, J. B. (1976) Role of maternal antibody causing immunosuppressive delay in the onset of plaque-forming cell responses in rats and rabbits. *See* Hemmings (1976), pp. 273–89.

Levine, B. B., Siraganian, G. F. and Schenkein, I. (1973) Allergy to human seminal plasma. *New Eng. J. Med.* **288**: 894–6.

Levis, W. R., Whalen, J. J. and Sherins, R. J. (1976) Mixed cultures of sperm and leukocytes as a measure of histocompatibility in man. *Science* **191**: 302–4.

Li, T. S. and Behrman, S. J. (1970) The sperm- and seminal plasma-specific antigens of human semen. *Fertil. Steril.* **21**: 565–73.

Lin, T-M. and Halbert, S. P. (1976) Placental localization of human pregnancy-associated plasma proteins. *Science* **193**: 1249–52.

Linscott, W. D. (1970) Effect of cell surface antigen density on immunological enhancement. *Nature* **228**: 824–7.

Little, C. C. (1924) The genetics of tissue transplantation in mammals. *J. Cancer Res.* **8**: 75–95.

Loke, Y. W. (1978) *Immunology and Immunotherapy of the Human Foetal–Maternal Interaction.* Elsevier/North-Holland.

Loke, Y. W., Cater, D. B., Whyte, A., Brook, S. S. and Day, S. (1980) Electrophoretic mobilities of human trophoblast cells and lymphocytes from pregnant women. *J. reprod. Immunol.* **2:** 45–52.

Lopo, A. C. and Vacquier, V. D. (1980) Sperm-specific surface antigenicity common to seven animal phyla. *Nature* **288**: 397–9.

Lord, E. M., Sensabaugh, G. F. and Stites, D. P. (1977) Immunosuppressive activity of human seminal plasma. I. Inhibition of *in vitro* lymphocyte activation. *J. Immunol.* **118**: 1704–11.

Mancini, R. E. (1971) Immunological approaches to fertility control, in *Control of Human Fertility*, Proc. 15th Nobel Symp. 1971, Diczfalusy, E. and Borell, U. (eds.), Wiley/Almqvist and Wigsell, pp. 157–78.

Mancini, R. E., Andrada, J. A., Saraceni, A., Bachmann, A. E., Lavieri, J. C. and Nemirovsky, M. (1965) Immunological and testicular response in man sensitized with human testicular homogenate. *J. clin. Endoc. Metab.* **25**: 859–75.

Marcus, Z., Freisheim, J., Herman, J. H. and Hess, E. V. (1977) *In vitro* cell-mediated immunity (CMI) to human seminal plasma fractions. *Fedn. Proc.* **36**: 371A.

Marcus, Z. and Hess, E. V. (1980) Local control of the immune response in the genital tract. *Fedn. Proc.* **39**: 459.

Maroni, E. S. and Parrott, D. M. V. (1973) Progressive increase in cell-mediated immunity against paternal transplantation antigens in parous mice after multiple pregnancies. *Clin. exp. Immunol.* **13**: 253–62.

Marti, J. J. and Herrman, U. (1977) Immunogestosis: a new etiologic concept of "essential" EPH gestosis, with special consideration of the primigravid patient. *Amer. J. Obstet. Gynecol.* **128**: 489–93.

Martin, B. J., Spicer, S. S. and Smythe, N. M. (1974) Cytochemical studies of the maternal surface of the syncytiotrophoblast of human early and term placentae. *Anat. Rec.* **178**: 769–86.

Maruta, H. and Moyer, D. L. (1965) Phagocytosis of spermatozoa in immunized and non-immunized guinea pigs. *Fed. Proc.* **24**: 450.

Marx, J. L. (1980) Natural killer cells help defend the body. *Science* **210**: 624–6.

Masson, P. L., Delire, M. and Cambiaso, C. L. (1977) Circulating immune complexes in normal human pregnancy. *Nature* **266**: 542–3.

Matangkasombut, P., Wattanasak, K., Sarntivijai, S. and Athavichitjanyaraks, M. (1979) Control mechanisms in materno-foetal immunological interactions. *See* Talwar (1979), pp. 525–34.

McIntyre, J. A. and Faulk, W. P. (1979*a*) Antigens of human trophoblast. Effects of heterologous anti-trophoblast sera on lymphocyte responses *in vitro*. *J. exp. Med.* **149**: 824–36.

McIntyre, J. A. and Faulk, W. P. (1979*b*) Trophoblast modulation of maternal allogeneic recognition. *Proc. nat. Acad. Sci.* **76**: 4029–32.

McLaren, A. (1975) Antigenic disparity: does it affect placental size, implantation, or population genetics? *See* Edwards *et al.* (1975), pp. 255–73.

McLean, J. M. and Scothorne, R. J. (1972) The fate of skin allografts in the rabbit uterus. *J. Anat.* **112**: 423–32.

McLean, J. M., Shaya, E. I. and Gibbs, A. C. C. (1980) Immune response to first mating in the female rat. *J. reprod. Immunol.* **1**: 285–95.

McManus, L. M., Naughton, M. A. and Martinez-Hernandez, A. (1976) Human chorionic gonadotropin in human neoplastic cells. *Cancer Res.* **36**: 3476–81.

Medawar, P. B. and Sparrow, E. M. (1956) The effects of adrenocortical hormones, adrenocorticotrophic hormone and pregnancy on skin transplantation immunity in mice. *J. Endocrinol.* **14**: 246–56.

Mendelsohn, J., Multer, M. M. and Bernheim, J. L. (1977) Inhibition of human lymphocyte stimulation by steroid hormones: cytokinetic mechanisms. *Clin. exp. Immunol.* **27**: 127–34.

Menge, A. C., Burkons, D. M. and Friedlaender, G. E. (1972) Occurrence of embryo mortality in rabbits isoimmunized against semen. *Int. J. Fertil.* **17**: 93–6.

Menge, A. C. and Lieberman, M. E. (1974) Antifertility effects of immunoglobulins from uterine fluids of semen-immunized rabbits. *Biol. Reprod.* **10**: 422–8.

Menge, A. C. and Protzman, W. P. (1967) Origin of the antigens in rabbit semen which induce antifertility antibodies. *J. Reprod. Fertil.* **13**: 31–40.

Mettler, L. and Schirwani, D. (1975) Macrophage migration inhibitory factor in female sterility. *Amer. J. Obstet. Gynecol.* **121**: 117–20.

Millette, C. F. and Bellvé, A. R. (1977) Temporal expression of membrane antigens during mouse spermatogenesis. *J. cell Biol.* **74**: 86–97.

Millette, C. F. and Bellvé, A. R. (1980) Selective partitioning of plasma membrane antigens during mouse spermatogenesis. *Developmental Biol.* **79**: 309–24.

Mizejewski, G. J. and Grimley, P. M. (1976) Abortogenic activity of antiserum to alpha-foetoprotein. *Nature* **259**: 222–4.

Monesi, V. (1972) Spermatogenesis and the spermatozoa. *See* Austin and Short (1972*a*), pp. 46–84.

Moore, H. D. M. (1981) Glycoprotein secretions of the epididymis in the rabbit and hamster: localization on epididymal spermatozoa and the effect of specific antibodies on fertilization *in vivo*. *J. exp. Zool.* **215**: 77–85.

Morgan, D. M. L. and Illei, G. (1980) Polyamine-polyamine oxidase interaction: part of maternal protective mechanism against fetal rejection. *Brit. med. J.* **280**: 1295–7.

Moudgal, N. R., Jagannadha Rao, A. and Prahalada, S. (1974) Can hormone antibodies be used as a tool in fertility control? *J. Reprod. Fertil. Suppl.* **21**: 105–23.

Muggleton-Harris, A. L. and Johnson, M. H. (1976) The nature and distribution of serologically detectable alloantigens on the preimplantation mouse embryo. *J. Embryol. exp. Morphol.* **35**: 59–72.

Muir, V. Y. and Turk, J. L. (1979) Immunological unresponsiveness during induction of experimental autoimmune orchitis in guinea-pigs: studies *in vivo* and *in vitro*. *Immunol.* **36**: 95–102.

Murgita, R. A., Anderson, I. C., Sherman, M. S., Bennich, H. and Wigzell, H. (1978) Effects of human alpha-foetoprotein in human B and T lymphocyte proliferation *in vitro*. *Clin. exp. Immunol.* **33**: 347–56.

Murgita, R. A., Goidl, E. A., Kontiainen, S. and Wigzell, H. (1977) α-fetoprotein induces suppressor T cells *in vitro*. *Nature* **267**: 257–9.

Murgita, R. A. and Tomasi, T. B. (1975*a*) Suppression of the immune response by α-fetoprotein. I. The effect of mouse α-fetoprotein on the primary and secondary antibody response. *J. exp. Med.* **141**: 269–86.

Murgita, R. A. and Tomasi, T. B. (1975*b*) Suppression of the immune response by α-fetoprotein. II. The effect of mouse α-fetoprotein on mixed lymphocyte reactivity and mitogen-induced lymphocyte transformation. *J. exp. Med.* **141**: 440–52.

Myles, D. G., Primakoff, P. and Bellvé, A. R. (1981) Surface domains of the guinea pig sperm defined with monoclonal antibodies. *Cell* **23**: 433–9.

Nelson, J. H. and Hall, J. E. (1964) Studies on the thymolymphatic system in humans. I. Morphologic changes in lymph nodes in pregnancy at term. *Amer. J. Obstet, Gynecol.* **90**: 482–4.

Newman, M. J. and Hines, H. C. (1980) Stimulation of maternal anti-lymphocyte antibodies by first gestation bovine fetuses. *J. Reprod. Fertil.* **60**: 237–41.

Noonan, F. P., Halliday, W. J., Morton, H. and Clunie, G. J. A. (1979) Early pregnancy factor is immunosuppressive. *Nature* **278**: 649–51.

Ockleford, C. D. and Whyte, A. (1977) Differentiated region of human placental cell surface associated with exchange of materials between maternal and foetal blood; coated vesicles. *J. Cell Sci.* **25**: 293–312.

Ogra, S. S., Weintraub, D. and Ogra, P. L. (1977) Immunologic aspects of human colostrum and milk. III. Fate and absorption of cellular and soluble components in the gastrointestinal tract of the newborn. *J. Immunol.* **119**: 245–8.

Ohno, S., Matsunaga, T., Epplen, J. T. and Hozumi, T. (1980) Interaction of viruses and lymphocytes in evolution, differentiation and oncogenesis. *See* Fougereau and Dausset (1980), pp. 577–98.

Oldstone, M. B. A., Tishon, A. and Moretta, L. (1977) Active thymus derived suppressor lymphocytes in human cord blood. *Nature* **269**: 333–5.

O'Rand, M. G. and Romrell, L. J. (1977) Appearance of cell surface auto- and isoantigens during spermatogenesis in the rabbit. *Developmental Biol.* **55**: 347–58.

O'Rand, M. G. and Romrell, L. J. (1980) Appearance of regional surface autoantigens during spermatogenesis: comparison of anti-testis and anti-sperm autoantisera. *Developmental Biol.* **75**: 431–41.

Orsini, F. and Shulman, S. (1971) The antigens and autoantigens of the seminal vesicle. I. Immunochemical studies on guinea pig vesicular fluid. *J. exp. Med.* **134**: 120–40.

Pachi, A., D'Amelio, R., Palmisano, L., Bilotta, P., Milano, C. F., De Gado, F. and Aiuti, F.

(1980) Immune complexes in the sera and amniotic fluids of human pregnancy. *J. perinat. Med.* **8**: 109–111.

Padma, M. C. (1972) Presence of a spermagglutinating factor in the normal serum of rabbits against homologous spermatozoa. *J. Reprod. Fertil.* **31**: 119–22.

Palm, J. (1974) Maternal-fetal histoincompatibility in rats: an escape from adversity. *Cancer Res.* **34**: 2061–5.

Parish, W. E., Carron-Brown, J. A. and Richards, C. B. (1967) The detection of antibodies to spermatozoa and to blood group antigens in cervical mucus. *J. Reprod. Fertil.* **13**: 469–83.

Parmely, M. J., Beer, A. E. and Billingham, R. E. (1976) *In vitro* studies on the T-lymphocyte population of human milk. *J. exp. Med.* **144**: 358–70.

Parr, E. L., Blanden, R. V. and Tulsi, R. S. (1980) Epithelium of mouse yolk sac placenta lacks H-2 complex alloantigens. *J. exp. Med.* **152**: 945–55.

Pavia, C., Siiteri, P. K., Perlman, J. D. and Stites, D. P. (1979) Suppression of murine allogeneic cell interactions by sex hormones. *J. reprod. Immunol.* **1**: 33–8.

Pavia, C. S. and Stites, D. P. (1980) Trophoblast regulation of maternal–paternal lymphocyte interactions. *Cell. Immunol.* **58**: 202–8.

Peck, A. B., Murgita, R. A. and Wigzell, H. (1978*a*) Cellular and genetic restrictions in the immunoregulatory activity of alpha-fetoprotein. I. Selective inhibition of anti-Ia-associated proliferative reactions. *J. exp. Med.* **147**: 667–83.

Peck, A. B., Murgita, R. A. and Wigzell, H. (1978*b*) Cellular and genetic restrictions in the immunoregulatory activity of alpha-fetoprotein. II. Alpha-fetoprotein-induced suppression of cytotoxic lymphocyte development. *J. exp. Med.* **148**: 360–72.

Pence, H., Petty, W. M. and Rocklin, R. E. (1975) Suppression of maternal responsiveness to paternal antigens by maternal plasma. *J. Immunol.* **114**: 525–8.

Pernis, B. and Vogel, H. J. (eds.) (1980) *Regulatory T Lymphocytes.* Academic Press.

Perry, J. S. (1981) The mammalian fetal membranes. *J. Reprod. Fertil.* **62**: 321–35.

Petersen, B. H., Lammel, J., Stites, D. P. and Brooks, G. F. (1980) Human seminal plasma inhibition of complement. *J. lab. clin. Med.* **96**: 582–91.

Pitcher-Wilmott, R. W., Hindocha, P. and Wood, C. B. S. (1980) The placental transfer of immunoglobulin G subclasses in human pregnancy. *Clin. exp. Immunol.* **41**: 303–8.

Pittard, W. B. and Bill, K. (1979) Immunoregulation by breast milk cells. *Cell. Immunol.* **42**: 437–41.

Playfair, J. H. L. (1979) *Immunology at a Glance.* Blackwell.

Ploegh, H. L., Orr, H. T. and Strominger, J. L. (1981) Major histocompatibility antigens: the human (HLA-A, -B, -C) and murine (H-2K, H-2D) Class I molecules. *Cell* **24**: 287–99.

Prehn, R. T. (1960) Specific homograft tolerance induced by successive matings and implications concerning choriocarcinoma. *J. nat. Cancer Inst.* **25**: 883–6.

Pruslin, F. H., Romani, M. and Rodman, T. C. (1979) Immunologic cross reactivity among mammalian sperm basic proteins. *J. cell Biol.* **83**: 219a.

Race, R. R. and Sanger, R. (1975) *Blood Groups in Man.* (6th edn.) Blackwell.

Radu, I. and Voisin, G. A. (1975) Ontogenesis of spermatozoa autoantigens in guinea pigs. *Differentiation* **3**: 107–14.

Ralph, P., Nakoinz, I. and Cohn, M. (1972) IgM-IgG1, 2, 3 relationship during pregnancy. *Nature* **238**: 344–5.

Rayfield, L. S., Brent, L. and Rodeck, C. H. (1980) Development of cell-mediated lympholysis in human fetal blood lymphocytes. *Clin. exp. Immunol.* **42**: 561–70.

Rebello, R., Green, F. H. Y. and Fox, H. (1975) A study of the secretory immune system of the female genital tract. *Brit. J. Obstet. Gynecol.* **82**: 812–6.

Redman, C. W. G. (1980) Immunological aspects of eclampsia and pre-eclampsia. *See* Hearn (1980), pp. 83–103.

Reid, B. L. (1965) Interaction between homologous sperm and somatic cells of the uterus and peritoneum in the mouse. *Exp. Cell Res.* **40**: 679–83.

Reid, B. L. and Blackwell, P. M. (1967) Evidence for the possibility of nuclear uptake of polymerised deoxyribonucleic acid of sperm phagocytosed by macrophages. *Aust. J. exp. Biol. med. Sci.* **45**: 323–6.

Reinherz, E. L. and Schlossman, S. F. (1980) The differentiation and function of human T lymphocytes. *Cell* **19**: 821–7.

Renshaw, H. W. and Gilmore, R. J. (1980) Alternative and classical complement pathways activity in sera from colostrum-fed and colostrum-deprived neonatal pigs. *Immunol.* **41**: 203–29.

Richards, R. C. (1979) Milk secretion: with special reference to its content of hormones, enzymes, and immunoreactive components. *Oxford Rev. Reprod. Biol.* **1**: 262–82.

Riera, C., Yantorno, C. and Shulman, S. (1967) Antigenic specificity of seminal plasma and the formation of autoantibodies. *Fedn. Proc.* **26**: 532.

Roberts, T. K. and Boettcher, B. (1969) Identification of human sperm-coating antigen. *J. Reprod. Fertil.* **18**: 347–50.

Robertson, M. (1979) Recognition, restriction and immunity. *Nature* **280**: 192–3.

Robertson, M. (1980) The life of a B lymphocyte. *Nature* **283**: 332–2.

Robertson, M. (1981) Genes of lymphocytes I: diverse means to antibody diversity. *Nature* **290**: 625–7.

Rocklin, R. E., Kitzmiller, J. L., Carpenter, C. B., Garovoy, M. R. and David, J. R. (1976) Maternal-fetal relation: absence of an immunologic blocking factor from the serum of women with chronic abortions. *New Eng. J. Med.* **295**: 1209–13.

Rocklin, R. E., Zuckerman, J. E., Alpert, E. and David, J. R. (1973) Effect of multiparity on human maternal hypersensitivity to foetal antigen. *Nature* **241**: 130–1.

Rodewald, R. (1973) Intestinal transport of antibodies in the newborn rat. *J. cell Biol.* **58**: 189–211.

Roitt, I. M. (1980) *Essential Immunology.* (4th edn.) Blackwell.

Rosenthal, A. S., Barcinski, M. A. and Rosenwasser, L. J. (1978) Function of macrophages in genetic control of immune responsiveness. *Fedn. Proc.* **37**: 79–85.

Rosenthal, M. (1977) Enhanced phagocytosis of immune complexes in pregnancy. *Clin. exp. Immunol.* **28**: 189–91.

Rümke, P. (1974) Autoantibodies against spermatozoa in infertile men. *J. Reprod. Fertil. Suppl.* **21**: 169–80.

Rümke, P. (1978) Autoantibodies against spermatozoa. *See* Cohen and Hendry (1978), pp. 67–79.

Rümke, P. (1980) Auto- and isoimmune reactions to antigens of the gonads and genital tract. *See* Fougereau and Dausset (1980), pp. 1065–92.

Russo, J. and Metz, C. B. (1974) The ultrastructural lesions induced by antibody and complement in rabbit spermatozoa. *Biol. Reprod.* **10**: 293–308.

Sacco, A. G. (1977) Antigenicity of the human zona pellucida. *Biol. Reprod.* **16**: 158–63.

Saji, F., Nakamuro, K., Wakao, T., Negoro, T., Tsuzuku, O. and Kurachi, K. (1977) Quantitative estimation of alloantisera against mouse H-antigens on motility of spermatozoa. *See* Boettcher (1977), pp. 215–20.

Saji, F., Nakamuro, K., Wakao, T. and Negoro, T. (1980) Sensitized T lymphocytes against paternal histocompatibility antigens cause intrauterine fetal death and growth retardation. *Acta obst. gynaecol. Jpn.* **32**: 1853–8.

Salinas, F. A., Silver, H. K. B., Sheikh, K. M. and Chandor, S. B. (1978) Natural occurrence of human tumor-associated anti-fetal antibodies during normal pregnancy. *Cancer* **42** (suppl.): 1653–9.

Schiff, R. I., Mercier, D. and Buckley, R. H. (1975) Inability of gestational hormones to account for the inhibitory effects of pregnancy plasmas on lymphocyte responses *in vitro*. *Cell. Immunol.* **20**: 69–80.

Schwimmer, W. B., Ustay, K. A. and Behrman, S. J. (1967) Sperm-agglutinating antibodies and decreased fertility in prostitutes. *Obstet. Gynecol.* **30**: 192–200.

Scott, J. S. and Jones, W. R. (eds.) (1976) *Immunology of Human Reproduction.* Academic Press/Grune and Stratton.

Searle, R. F., Johnson, M. H., Billington, W. D., Elson, J. and Clutterbuck-Jackson, S. (1974) Investigation of H-2 and non-H2 antigens on the mouse blastocyst. *Transpl.* **18**: 136–41.

Searle, R. F., Sellens, M. H., Elson, J., Jenkinson, E. J. and Billington, W. D. (1976) Detection of alloantigens during preimplantation development and early trophoblast differentiation in the mouse by immunoperoxidase labelling. *J. exp. Med.* **143**: 348–59.

Seelig, L. L. (1980) Dynamics of leukocytes in rat mammary epithelium during pregnancy and lactation. *Biol. Reprod.* **22**: 1211–7.

Sellens, M. H. and Jenkinson, E. J. (1975) Permeability of the mouse zona pellucida to immunoglobulins. *J. Reprod. Fertil.* **42**: 153–7.

Sellens, M. H., Jenkinson, E. J. and Billington, W. D. (1978) Major histocompatibility complex and non-major histocompatibility complex antigens on mouse ectoplacental cone and placental trophoblastic cells. *Transpl.* **25**: 173–9.

Sethi, K. K. and Brandis, H. (1980) IgG Fc-binding receptors on spermatozoa. *Eur. J. Immunol.* **10**: 964–5.

Shacter, B., Muir, A., Gyves, M. and Tasin, M. (1979) HLA-A, B compatibility in parents of offspring with neural-tube defects or couples experiencing involuntary foetal wastage. *Lancet* 1979/1: 796–9.

Shalev, A. (1980) Pregnancy-induced H-Y antibodies and their transmission to the foetus in rats. *Immunol.* **39**: 285–9.

Shivers, C. A. and Dunbar, B. S. (1977) Autoantibodies to zona pellucida: a possible cause for infertility in women. *Science* **197**: 1082–4.

Shulman, S. (1971) Antigenicity and autoimmunity in sexual reproduction: a review. *Clin. exp. Immunol.* **9**: 267–88.

Shulman, S. (1972) Immunologic barriers to fertility. *Obstet. gynecol. Survey* **27**: 553–606.

Shulman, S. (1974) *Tissue Specificity and Autoimmunity.* (Molec. Biol. Biochem. Biophys. 16.) Chapman and Hall/Springer-Verlag.

Shulman, S. (1978) Agglutinating and immobilizing antibodies to spermatozoa. *See* Cohen and Hendry (1978), pp. 81–99

Shulman, S. and Bronson, P. (1969) Immunochemical studies on human seminal plasma. II. The major antigens and their fractionation. *J. Reprod. Fertil.* **18**: 481–91.

Shulman, S., Patel, S. and Stamm, E. (1977) Antibodies to spermatozoa. VII. Humoral and cellular aspects of sperm immunity and infertility. *Reprod.* **3**: 125–34.

Shulman, S., Yantorno, C., Soanes, W. A., Gonder, M. J. and Witebsky, E. (1966) Studies on organ specificity. XVI. Urogenital tissues and autoantibodies. *Immunol.* **10**: 99–113.

Siiteri, P. K., Febres, F., Clemens, L. E., Chang, R. J., Gondos, B. and Stites, D. (1977) Progesterone and maintenance of pregnancy: is progesterone nature's immunosuppressant? *Ann. N.Y. Acad. Sci.* **286**: 384–97.

Slobodian, P. W., Carlson, G. A. and Wegmann, T. G. (1979) The processing of T-lymphocytes by the gut of the suckling neonate. *J. reprod. Immunol.* **1**: 23–31.

Smith, R. N. and Powell, A. E. (1974) The adoptive transfer of pregnancy-induced unresponsiveness to male skin grafts with thymus-dependent cells. *J. exp. Med.* **146**: 899–904.

Solomon, J. B. and Horton, J. D. (eds.) (1977) *Developmental Immunobiology.* (Procs. Symp. Developmental Immunobiology, Aberdeen, 1977.) Elsevier/North-Holland.

Stahn, R., Fabricius, H-Å. and Hartleiner, W. (1978) Suppression of human T-cell colony formation during pregnancy. *Nature* **276**: 831–2.

Stephens, S., Dolby, J. M., Montreuil, J. and Spik, G. (1980) Differences in inhibition of the growth of commensal and enteropathogenic strains of *E. coli* by lactotransferrin and secretory IgA isolated from human milk. *Immunol.* **41**: 597–603.

Steven, D. H. (ed.) (1975) *Comparative Placentation: Essays in Structure and Function.* Academic Press.

Steward, M. W. (1974) *Immunochemistry.* Chapman and Hall.

Stimson, W. H. (1980) Are pregnancy-associated serum proteins responsible for the inhibition of lymphocyte transformation by pregnancy serum? *Clin. exp. Immunol.* **40**: 157–60.

Stites, D. P. and Erickson, R. P. (1975) Suppressive effect of seminal plasma on leucocyte activation. *Nature* **253**: 727–9.

Stone, S. L., Huckle, W. R. and Oliphant, G. (1980) Identification and hormonal control of reproductive-tract-specific antigens present in rabbit oviductal fluid. *Gamete Res.* **3**: 169–77.

Streilein, J. W. and Klein, J. (1980) Neonatal tolerance of H-2 alloantigens. I. I region modulation of tolerogenic potential of K and D antigens. *Proc. roy. Soc B* **207**: 461–74.

Strelkauskas, A. J., Davies, I. J. and Dray, S. (1978) Longitudinal studies showing alterations in the levels and functional response of T and B lymphocytes in human pregnancy. *Clin. exp. Immunol.* **32**: 531–9.

Strominger, J. L. (1980) Structure of products of the major histocompatibility complex in man and mouse. *See* Fougereau and Dausset (1980), pp. 541–54.

Sundqvist, K-G., Bergström, S. and Håkansson, S. (1977) Surface antigens of human trophoblasts. *Dev. comp. Immunol.* **1**: 241–54.

Symons, D. B. A. and Herbert, J. (1971) Incidence of immunoglobulins in fluids of the rabbit genital tracts and the distribution of IgG-globulin in the tissues of the female tract. *J. Reprod. Fertil.* **24**: 55–62.

Talwar, G. P. (ed.) (1979) *Recent Advances in Reproduction and Regulation of Fertility.* Elsevier/North-Holland.

Talwar, G. P. (1980) *Immunology of Contraception.* Edward Arnold.

Talwar, G. P., Manhar, S. K., Tandon, A., Ramakrishnan, S. and Gupta, S. K. (1980) Hormones and antibodies in the study and control of reproductive processes. *See* Fougereau and Dausset (1980), pp. 1118–36.

Talwar, G. P., Naz, R. K., Das, C. and Das, R. P. (1979) A practicable immunological approach to block spermatogenesis without loss of androgens. *Proc. nat. Acad. Sci.* **76**: 5882–5.

Tarter, T. H. and Anderson, D. J. (1980) Immunosuppressive activities of accessory sex organ fluids in the mouse. *Fedn. Proc.* **39**: 458.

Tauber, P. F., Zaneveld, L. J. D., Propping, D. and Schumacher, G. F. B. (1975) Components of human split ejaculates. I. Spermatozoa, fructose, immunoglobulins, albumin, lactoferrin, transferrin and other plasma proteins. *J. Reprod. Fertil.* **43**: 249–67.

Taussig, M. J. (1980) Antigen-specific T-cell factors. *Immunol.* **41**: 759–87.

Taylor, N. J. (1980) Sperm populations in the female genital tract of rabbits are not immunologically different from the ejaculate. Paper presented at Soc. for the Study of Fertility Winter Meeting, December 1980.

Taylor, P. V. and Hancock, K. W. (1975) Antigenicity of trophoblast and possible antigen-masking effects during pregnancy. *Immunol.* **28**: 973–82.

Tekelioğlu-Uysal, M., Edwards, R. G. and Kişnişçi, H. A. (1975) Ultrastructural relationships between decidua, trophoblast and lymphocytes at the beginning of human pregnancy. *J. Reprod. Fertil.* **42**: 431–8.

Tiilikainen, A., Schroder, J. and de la Chapelle, A. (1974) Fetal leucocytes in the maternal circulation after delivery. II. Masking of HL-A antigens. *Transpl.* **17**: 355–60.

Tinneberg, H., Birke, R. and Mettler, L. (1980) Effect on fertility of female mice of the transfer of lymphocytes from females previously immunized with mouse spermatozoa. *J. Reprod. Fertil.* **58**: 469–73.

Toullet, F. and Voisin, G. A. (1976) Passive transfer of allergic aspermatogenic orchi-epididymitis (AIAO) by antispermatozoal sera. *Clin. exp. Immunol.* **26**: 549–62.

Touraine, J. L., Traeger, J., Betuel, H., Brochier, H., Brochier, J., Dubernard, J. M., Revillard, J. P. and Triau, R. (eds.) (1980) *Transplantation & Clinical Immunology* 11. Symp. Fondation Merieux vol. 3. Excerpta Medica.

Tovey, L. A. D. and Maroni, E. S. (1976) Rhesus isoimmunisation. *See* Scott and Jones (1976), pp. 187–227.

Tuffrey, M., Bishun, N. P. and Barnes, R. D. (1969) Porosity of the mouse placenta to maternal cells. *Nature* **221**: 1029–30.

Tung, K. S. K. (1977) The nature of antigens and pathogenetic mechanisms in auto-immunity to sperm. *See* Edidin and Johnson (1977), pp. 157–85.

Tung, K. S. K. (1980) Autoimmunity of the testis, in *Immunological Aspects of Infertility and Fertility Regulation*, Dhindsa, D. S. and Schumacher, G. F. B. (eds.). North-Holland, pp. 33–92.

Uhr, J. W., Capra, J. D., Vitetta, E. S. and Cook, R. G. (1979) Organization of the Immune Response genes. *Science* **206**: 292–7.

Vaerman, J-P. and Férin, J. (1974) Local immunological response in the vagina, cervix, and endometrium. *See* Diczfalusy (1974), pp. 281–305.

Vallotton, M. B. and Forbes, A. P. (1966) Antibodies to cytoplasm of ova. *Lancet* 1966/2: 264–5.

Voisin, G. A. (1979) Rejection and facilitation immune reactions of the mother towards the father-inherited antigens of the conceptus. *See* Talwar (1979), pp. 515–24.

Voisin, G. A. (1980) Humoral and cellular suppressor mechanisms involved in the acceptance of the fetal allograft. *See* Touraine *et al.* (1980), pp. 129–36.

Voisin, G. A. and Chaouat, G. (1974) Demonstration, nature and properties of maternal antibodies fixed on placenta and directed against paternal antigens. *J. Reprod. Fertil. Suppl.* **21**: 89–103.

Voisin, G. A., Toullet, F. and d'Almeida, M. (1974) Characterization of spermatozoal auto-, iso- and allo-antigens. *See* Diczfalusy (1974), pp. 173–201.

Vojtíšková, M. and Pokorná, Z. (1972) Cellular antigens of mouse spermatozoa as possible markers of gene action, in *The Genetics of the Spermatozoon*, Beatty, R. A. and Gluecksohn-Waelsch, S. (eds.). University of Edinburgh Symposium, pp. 160–76.

von Schoultz, B., Stigbrand, T. and Tarnvik, A. (1973) Inhibition of PHA-induced lymphocyte stimulation by the Pregnancy Zone Protein. *FEBS Lett.* **38**: 23–6.

Wachtel, S. S. and Silvers, W. K. (1971) Skin homografts: tolerogenic versus immunogenic influences in mice. *J. exp. Med.* **133**: 921–37.

Walker, J. S., Freeman, C. B. and Harris, R. (1972) Lymphocyte reactivity in pregnancy. *Lancet* 1972/3: 469.

Wass, M., McCann, K. and Bagshawe, K. D. (1975) Isolation of antibodies to HCG/LH from human sera. *Nature* **274**: 368–70.

Watson, D. L. (1980) Immunological functions of the mammary gland and its secretion—comparative review. *Aust. J. biol. Sci.* **33**: 403–22.

Wegmann, T. G., Singh, B. and Carlson, G. A. (1979*a*) Allogeneic placenta is a paternal strain antigen immunoabsorbent. *J. Immunol.* **122**: 270–4.

Wegmann, T. G., Waters, C. A., Drell, D. W. and Carlson, G. A. (1979*b*) Pregnant mice are not primed but can be primed to fetal alloantigens. *Proc. nat. Acad. Sci.* **76**: 2410–4.

Wegmann, T. G., Leigh, J. B., Carlson, G. A., Mosmann, T. R., Raghupathy, R. and Singh, B. (1980) Quantitation of the capacity of the mouse placenta to absorb monoclonal anti-fetal H-2K antibody. *J. reprod. Immunol.* **2**:53–9.

Weinstein, Y. (1977) 20α-hydroxysteroid dehydrogenase: a T lymphocyte-associated enzyme. *J. Immunol.* **119**: 1223–9.

Werthamer, S., Govindaraj, S. and Amaral, L. (1976) Placenta, transcortin, and localized immune response. *J. clin. Invest.* **57**: 1000–8.

Whyte, A. (1978) H.C.G. and trophoblast antigenicity. *Lancet* 1978/2:1003.

Whyte, A. and Loke, Y. W. (1979) Antigens of the human trophoblast membrane. *Clin. exp. Immunol.* **37**: 359–66.

Wickings, E. J. and Nieschlag, E. (1980) Suppression of spermatogenesis over two years in rhesus monkeys actively immunized with follicle-stimulating hormone. *Fertil. Steril.* **34**: 269–74.

Wiley, L. M. (1980) Early mouse embryonic cell surface antigens that are detectable by antisera to human chorionic gonadotropin. *Exp. Cell Res.* **129**: 47–54.

Winchester, R. J., Fu, S. M., Wernet, P., Kunkel, H. G., Dupont, B. and Jersild, C. (1975) Recognition by pregnancy serums of non-HL-A alloantigens selectively expressed on B lymphocytes. *J. exp. Med.* **141**: 924–9.

Wira, C. R. and Sandoe, C. P. (1980) Hormonal regulation of immunoglobulins: influence of estradiol on immunoglobulins A and G in the rat uterus. *Endocrinol.* **106**: 1020–6.

Witkin, S. S., Good, R. A. and Day, N. K. (1980) Presence of Fc receptors on spermatozoa and their utilization in a radioimmunoassay and enzyme-linked immunoassay for the detection of circulating immune complexes in human serum. *Fedn. Proc.* **39**: 673.

Wolf, R. L. (1977) An inhibitor of lymphocyte proliferation and lymphokine production released by unstimulated foetal monocytes. *Clin. exp. Immunol.* **27**: 464–8.

Wood, G., Reynard, J., Krishnan, E. and Racela, L. (1978*a*): Immunobiology of the human placenta. I. IgGFc receptors in trophoblastic villi. *Cell. Immunol.* **35**: 191–204.

Wood, G., Reynard, J., Krishnan, E. and Racela, L. (1978*b*) Immunobiology of the human placenta. II. Localization of macrophages, *in vivo* bound IgG and C3. *Cell. Immunol.* **35**: 205–16.

Wyle, F. A. and Kent, J. R. (1977) Immunosuppression by sex hormones. I. The effect upon PHA- and PPD-stimulated lymphocytes. *Clin. exp. Immunol.* **27**: 407–15.

Wynn, R. M. (1969) Noncellular components of the placenta. *Amer. J. Obstet. Gynecol.* **103**: 723–39.

Yachnin, S. (1975) Fetuin, an inhibitor of lymphocyte transformation. The interaction of fetuin with phytomitogens and a possible role for fetuin in fetal development. *J. exp. Med.* **141**: 242–56.

Zimmerman, E. F., Voorting-Hawking, M. and Michael, J. G. (1977) Immunosuppression by mouse sialylated α-foetoprotein. *Nature* **265**: 354–6.

Index